ISO 14001 AUDITING MANUAL

Gayle Woodside

Patrick Aurrichio

McGraw-Hill
New York • San Francisco • Washington, D.C. • Auckland
Bogotá • Caracas • Lisbon • London • Madrid • Mexico City
Milan • Montreal • New Delhi • San Juan • Singapore
Sydney • Tokyo • Toronto

658.408
W89 is

Library of Congress Cataloging-in-Publication Data

Woodside, Gayle
   ISO 14001 auditing manual / Gayle Woodside, Patrick Aurrichio.
     p. cm.
   ISBN 0-07-134907-3
    1. ISO 14000 Series Standards—Auditing—Handbooks, manuals, etc.
   2. Compliance auditing—Handbooks, mauals, etc.   I. Aurrichio, Patrick.
   II. Title.
   TS155.7.W64 1999
   658.4'08—dc21   99-046588
                                                         CIP

McGraw-Hill

A Division of The McGraw·Hill Companies

1 2 3 4 5 6 7 8 9 0 DOC/DOC 9 0 9 8 7 6 5 4 3 2 1 0 9

ISBN 0-07-134907-3

The sponsoring editor for this book was Bob Esposito, the editing supervisor was Peggy Lamb, and the production supervisor was Sherri Souffrance. It was set in Fairfield by McGraw-Hill's Professional Book Group composition unit in Hightstown, New Jersey, in cooperation with Spring Point Publishing Services.

Printed and bound by R. R. Donnelley & Sons Company.

This book is printed on recycled, acid-free paper containing a minimum of 50% recycled, de-inked fiber.

McGraw-Hill books are available at special quantity discounts to use as premiums and sales promotions, or for use in corporate training programs. For more information, please write to the Director of Special Sales, McGraw-Hill, 11 West 19th Street, New York, NY 10011. Or contact your local bookstore.

CONTENTS

v

PART TWO. GUIDANCE FOR PLANNING AND CONDUCTING AN EMS AUDIT

PART THREE. PUTTING IT ALL TOGETHER

PREFACE

ISO 14001—the Environmental Management Systems standard—was published in final form in September 1996, and is widely considered to be the most important standard in ISO's environmental management series. The standard is a process standard—that is, it specifies a management process and not an end goal. It sets forth a framework for an environmental management system (EMS) which includes numerous substantive elements such as an environmental policy defined by top management; identification of significant environmental aspects; the setting of environmental objectives and targets; clear identification and communication of roles and responsibilities within the EMS; the establishment of procedures to ensure operational control of those activities which could impact the environment; and a means of checking the organization's EMS with respect to the requirements of ISO 14001. The crafters of ISO 14001 were careful to ensure that it is applicable to organizations of varying sizes and circumstances, including small, medium, and large enterprises. The intent of the standard is to drive environmental improvements worldwide through a systematic approach to environmental management.

Now, some three years later after ISO 14001 was finalized, organizations are finding that ISO 14001 is living up to its lofty expectations. Benefits from implementing the standard are numerous and include: (1) an EMS aligned with ISO 14001 makes the task of managing environmental matters "system dependent" rather than "person dependent"; (2) employees (from top to bottom!), as well as on-site contractors, who did not traditionally see themselves as needing to be involved with the environmental management process, now become fully integrated into the EMS and understand their role in supporting it; and (3) the setting of environmental objectives and targets is based on significant environmental aspects and impacts—this goes beyond simply relying on legal/regulatory requirements, and allows the flexibility to include additional areas for environmental improvement.

In addition to the above, an organization will reap substantial benefits from its EMS audit—a fundamental component of the ISO 14001 standard. This audit not only verifies that the organization is implementing its EMS according to planned arrangements, but offers opportunities to identify areas for continual improvement. Further, the EMS audit fosters employee awareness and provides valuable input for top management to assess the suitability, adequacy, and effectiveness of the EMS.

This book presents the reader with complete information necessary to develop and conduct an EMS audit. Topics include audit protocol, auditor and auditee responsibilities, and conducting the EMS audit. We have provided templates, checklists, and tables to add clarity to our comments and suggestions. We hope that you, the reader and auditor, find this book both practical and thought-provoking.

—*Gayle Woodside*
Patrick Aurrichio

INTRODUCTION TO EMS AUDITING

INTRODUCTION TO ISO 14001

ENVIRONMENTAL MANAGEMENT SYSTEMS SPECIFICATION—A GENERAL OVERVIEW

ISO 14001, which was finalized and issued as a first edition September 1, 1996, is the most widely recognized environmental management system standard. It is a specification standard, which means that organizations that conform to its requirements can become registered to the standard. (*Note:* Outside the United States, the term *certified* is used instead of *registered*; although the two terms are synonymous, the authors will use the term *registered* throughout the book for consistency.)

ISO 14001 was written as a consensus standard with nearly 50 countries participating in its development and over 100 countries endorsing it as an international standard. The standard is applicable to all types and sizes of organizations, and it accommodates diverse geographical, cultural, and social conditions. It can be applied to all parts or any single part of an organization and/or its activities, products, and services. The ISO 14001 standard is organized as follows:

- *Introduction*   This nonmandatory section sets the tone and context for implementation of the standard. In addition to background information about applicability and use of the standard, it provides a very basic model of the framework that the standard sets, which includes the five major elements and the concept of continual improvement. This model is presented in Figure 1-1.

- *Scope (Section 1)*   This section specifies applicability and introduces Annex A as nonmandatory guidance.

- *Normative references (Section 2)*   None are listed at present.

- *Definitions (Section 3)*   Thirteen definitions that apply to the standard are presented. Examples include definitions of environmental aspect, continual improvement, environmental performance, and interested party.

- *Environmental management system (EMS) requirements (Section 4)*   This section details the requirements of the EMS to which an organization must conform if it wants to become registered to the standard.

- *Annex A*   This nonmandatory (informative) annex provides guidance on the EMS requirements section so that misinterpretation of the standard can be avoided.

- *Annex B*   This second nonmandatory (informative) annex provides two tables that identify the links and broad technical correspondences between ISO 14001 and ISO 9001.

- *Annex C*   This final nonmandatory (informative) annex provides a bibliography of ISO 9000 standards, environmental audit guidelines, and an EMS guidance document.

There are five major elements of the standard, as shown in Figure 1-1—policy, planning, implementation, and operation; checking and corrective action; and management review. These elements interact with each other to form the framework of an integrated, systematized approach to environmental management, with the result being continual improvement of the overall system and, ultimately, environmental performance.

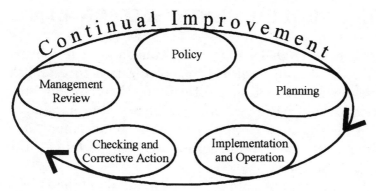

FIGURE 1-1.   ISO 14001 model.

GENERAL REQUIREMENTS (SECTION 4.1)

Section 4.1 of ISO 14001 is brief and to the point. Essentially, the section requires an organization to establish and maintain an EMS that conforms to the requirements detailed throughout the rest of Section 4—the mandatory sections of the standard. To understand this requirement, it is appropriate to review several definitions that pertain to the standard and the EMS.

First, what is an *organization?* The standard defines it as a "company, corporation, firm, authority or institution, or part or combination thereof, whether incorporated or not, public or private, that has its own function and administration" (ISO 14001, Section 3.12). This very broad and flexible definition essentially allows many organizational entities to align their EMSs with the requirements of ISO 14001.

Another definition that is integral to the standard is *environmental management system* (ISO 14001, Section 3.5). The ISO 14001 standard defines this term as "the part of the overall management system that includes organizational structure, planning activities, responsibilities, practices, procedures, processes, and resources for developing, implementing, achieving, reviewing, and maintaining the environmental policy."

Further, the term *continual improvement* (ISO 14001, Section 3.1) warrants some discussion at this point. The writers of the standard were careful in crafting the verbiage of this definition, which reads "process of enhancing the environmental management system to achieve improvements in overall environmental performance in line with the organization's environmental policy." Thus *continual improvement* is not improvement of environmental performance per se but rather improvement of the EMS, which then leads to improvement in environmental performance.

Although the improvement of environmental performance is not a prerequisite to obtaining registration to ISO 14001, implementation of the elements of the standard, as mentioned above, is likely to result in such improvements. The requirements follow the "plan, do, check, and act" model of the Shewharz cycle, which forms the basis of continual improvement of the system. The implementation of an EMS that conforms to ISO 14001 should allow the organization to accomplish the following management system objectives:

- Establish an environmental policy appropriate to the organization's activities, products, and services
- Identify significant environmental aspects and impacts
- Identify legal requirements
- Set environmental objectives and targets and the program to accomplish them
- Define structure and responsibility for elements of the EMS
- Involve employees at relevant levels in the EMS process
- Enhance communication with employees and external interested parties
- Schedule audits of the EMS
- Conduct management review of the EMS for determination of suitability, adequacy and effectiveness, and continual improvement of the EMS itself.

ENVIRONMENTAL POLICY (SECTION 4.2)

A well-structured EMS is predicated upon a strong environmental policy that is defined by top management. Anyone who has worked in the business world knows how important top management commitment is. Essentially, without it, the best-laid plans are often doomed for failure. The crafters of the ISO 14001 standard recognized this and made top management responsible for defining the organization's environmental policy. Indeed, many ISO 14001 experts think that this is the most prescriptive element of the standard. There are certain requirements that the environmental policy must meet, including the following:

- It must be appropriate to the nature, scale, and environmental impacts of the organization's activities, products, and services.

- It must include a commitment to continual improvement and prevention of pollution.

- It must include a commitment to comply with relevant environmental legislation and regulations and with other (including voluntary) requirements to which the organization subscribes.

- It must provide a framework for setting and reviewing environmental objectives and targets.

- It must be documented, implemented, and maintained.

- It must be communicated to all employees.

- It must be available to the public.

PLANNING (SECTION 4.3)

"Fail to plan and plan to fail" is an appropriate adage to keep in mind when setting up an EMS that conforms to ISO 14001. After defining the environmental policy, planning is the next important step in the ISO 14001 process.

ENVIRONMENTAL ASPECTS (SECTION 4.3.1)

The planning process begins with the identification of environmental aspects and, subsequently, significant environmental aspects. This concept of identifying environmental aspects is unique to the ISO 14001 standard, and although many organizations may have considered environmental outcomes when setting priorities, few are likely to have developed procedures for a formalized process, as required by the standard.

The term *environmental aspect* is defined as an "element of an organization's activities, products, or services that can interact with the environment." There is a note associated with the definition that defines a *significant environmental aspect* as one "that has or can have a significant environmental impact"

(Section 3.3). An *environmental impact* is defined as "any change to the environment, whether adverse or beneficial, wholly or partially resulting from an organization's activities, products, or services" (Section 3.4). Having provided definitions for environmental aspect and impact, the standard specifies that "the organization shall establish and maintain a procedure to identify environmental aspects of its activities, products or services that it can control or over which it can be expected to have an influence in order to determine those which have or can have significant impacts on the environment" (Section 4.3.1). This requirement brings the reader back to the note associated with *environmental aspect,* which, in essence, calls aspects with significant environmental impacts, *significant aspects.* It is important to note that the standard does not provide a definition for the term *significant.* Thus, the organization is able to use a variety of criteria to determine what is significant.

Although Annex A of ISO 14001 is for informative reference only (that is, its contents are guidance and not requirements), there is some useful information with respect to environmental aspects within it. Specifically, the annex suggests that "organizations should determine what their environmental aspects are, taking into account the inputs and outputs associated with their current and relevant past activities, products and/or services" (Section A.3.1).

This gives organizations a starting point. Annex A goes on to state, "The process is intended to identify significant environmental aspects associated with activities, products or services, and is not intended to require a detailed life cycle assessment. Organizations do not have to evaluate each product, component or raw material input. They may select categories of activities, products or services to identify those aspects most likely to have a significant impact." Thus, by beginning with broad-based aspects associated with categories of activities, products, and services, the process can be simplified and still be effective. Examples of types of inputs and outputs that the organization might want to consider include inputs and outputs from normal operating conditions, shutdown and startup conditions, and foreseeable (realistic) emergency situations.

LEGAL AND OTHER REQUIREMENTS (SECTION 4.3.2)

This element of ISO 14001 is succinct and to the point. It requires an organization to establish and maintain a procedure for identifying and providing access to legal requirements applicable to its activities, products, or services. Further, if the organization subscribes to other requirements—such as voluntary agency or industry requirements—these requirements must also be part of the procedure.

OBJECTIVES AND TARGETS (SECTION 4.3.3)

ISO 14001 requires an organization to establish and maintain documented objectives and targets. Elements to consider when establishing these objectives and targets include

- Relevant legal requirements and other requirements to which the organization subscribes
- Significant environmental aspects of the organization's activities, products, and services
- Technological options available to the organization
- The organization's financial, operational, and business requirements
- The views of interested parties

This seemingly short list actually covers a broad range of topics, and in addition to considering the elements above, the standard requires that the objectives and targets be consistent with the organization's environmental policy and with the commitment to prevention of pollution.

ENVIRONMENTAL MANAGEMENT PROGRAM (SECTION 4.3.4)

Once the organization sets its environmental objectives and targets, the ISO 14001 standard requires the organization to establish and maintain an environmental management program to achieve them. (Note: The use of the term *environ-*

mental management program in the standard relates singularly to a program for achieving objectives and targets and not to what is conventionally considered an environmental management program such as waste management program, air monitoring program, and others.) The program must include who, when, and how. In other words, the following must be defined:

- The designation of responsibility for achieving objectives and targets at each relevant function and level of the organization
- The time frame in which they are to be achieved
- The means by which they are to be achieved

The environmental management program must be updated as new developments or modifications to activities, products, and services warrant.

IMPLEMENTATION AND OPERATION (SECTION 4.4)

In the plan, do check, act model discussed above, this section is the "do." Essentially, the organization builds its EMS based on its environmental policy and planning elements. The longest section of the standard, this section encompasses structure and responsibility; training, awareness and competence; communication; EMS documentation; document control; operational control; and emergency preparedness and response.

STRUCTURE AND RESPONSIBILITY (SECTION 4.4.1)

Once again, the ISO 14001 standard is flexible in its approach when identifying requirements for structure and responsibility. Exactly who is responsible for the various duties that facilitate an effective EMS is left up to management to define, based on existing circumstances of the organization. There is no one "right" way to define roles, responsibility, and authority, but they must be defined to make the EMS system not person dependent, which is one of the many strengths inherent in ISO 14001. Requirements in this section include

- Roles, responsibilities, and authorities must be defined, documented, and communicated.
- Management must provide resources essential to the implementation and control of the EMS.
- Resources should include human resources and specialized skills, technology, and financial resources.
- Top management must appoint representative(s) to establish, implement, and maintain the EMS.
- Top management must appoint representative(s) to report on the performance of the EMS to them for review and as a basis for improvement of the EMS.

TRAINING, AWARENESS, AND COMPETENCE (SECTION 4.4.2)

Training of personnel is integral to proper functioning of the EMS. Because training depends on the organization's activities, products, and services, the organization must identify training needs. The ISO 14001 standard requires all personnel whose work may impact the environment to receive appropriate training. In addition, the organization must establish and maintain procedures to make employees or members at each relevant level aware of

- The importance of conformance with the environmental policy and procedures and with the requirements of the EMS
- The significant environmental impacts, actual or potential, or their work activities and the environmental benefits of improved personal performance
- Their roles and responsibilities in achieving conformance with the environmental policy and procedures and with the requirements of the environmental management sys-

tem, including emergency preparedness and response requirements

• The potential consequences of departure from specified operating procedures

Finally, the standard requires that personnel performing tasks that could cause a significant impact must be competent to perform that task, based on education, training, or experience.

COMMUNICATION (SECTION 4.4.3)

Communication is one of the most important elements of the EMS. The ISO 14001 standard requires organizations to establish and maintain procedures for both internal and external communication about the significant environmental aspects and the EMS. Internal communication is expected to be "multidirectional"—not just from the top down, but from the bottom up and throughout all relevant functions and levels of the organization. The organization must also have a documented process for receiving, documenting, and responding to relevant external communication. Further, the organization must consider processes for external communication about its significant aspects and record its decision about which, if any, it will deploy. Essentially, this requires the organization to consider processes for voluntary reporting to the public, although the organization is allowed to elect not to voluntarily report at all.

EMS DOCUMENTATION (SECTION 4.4.4)

The requirements for EMS documentation look deceptively simple. In essence, the standard requires the organization to establish and maintain information that describes the core elements of the EMS and their interaction and to provide direction to related documentation. The information can be in paper or electronic form.

This is actually a formidable task. In meeting these requirements, the organization must first define the core elements of the EMS, which are those documents that the organization relies upon to define the way it manages its entire environmental system. Certainly, the environmental policy would be considered a core element. Other documents that are part of the EMS, but that are not considered "core" and are related to the core documents, must also be listed or otherwise referenced.

Although not specifically required by the standard, many organizations are choosing to develop an EMS manual. This manual can describe in detail how the EMS system works and can introduce the documents that interact (i.e., supporting documents) with the core elements. In the ISO 9000 world—that of quality systems—management system documentation and document control requirements are very strictly defined. The drafters of ISO 14001 took special care not to be prescriptive with these requirements because they wanted the focus of the EMS standard to be on environmental management, not "paper management." Nonetheless, organizations that have mature EMSs might want to give consideration to fully documenting the system—beyond what is required in the standard—because it will give the various parts of the system consistency and effectiveness. This also supports the concept of developing an EMS that is system not person dependent.

DOCUMENT CONTROL (SECTION 4.4.5)

Under the ISO standard, documents are different from records and thus are controlled differently. Documents are procedures, manuscripts, forms, and other documentation that is relied upon to conduct current or planned activities. Records, on the other hand, show evidence that an event or activity occurred. Examples include meeting minutes, completed forms, and presentation materials.

Once the core elements of the EMS are defined, along with documents that are related to the core elements, these documents must be controlled. The standard requires the organization to establish and maintain procedures to ensure the following:

- Documents can be located.
- Documents are periodically reviewed, revised as necessary,

and approved for adequacy by authorized personnel.

- Current versions of the documents are available where needed.

- Obsolete documents are promptly removed or otherwise ensured against unintended use.

- Obsolete documents retained for legal and/or knowledge-preservation purposes are suitably identified.

- Responsibilities are established concerning the creation and modification of the various types of documents.

In addition, the documents must be legible, dated, readily identifiable, maintained in an orderly manner, and retained for a specified period of time.

OPERATIONAL CONTROL (SECTION 4.4.6)

Operational control involves all employees whose job functions have the potential to cause an impact on the environment. As such, these employees play a key role in the proper functioning of the EMS. As specified above, Section 4.4.2 of ISO 14001 requires that employees be made aware of the environmental impacts of their actions and potential impacts from deviating from established procedures. This section, then, describes how the procedures for day-to-day operations are established and maintained.

The first requirement of this part of the standard is for the organization to identify operations and activities—including maintenance—associated with significant environmental aspects and in line with the policy and objectives and targets. Once identified, the organization must then ensure that these operations and activities are carried out under specified conditions. To do that, the organization must

- Establish and maintain documented procedures to cover situations where their absence could lead to deviations from the environmental policy and objectives and targets

- Stipulate operating criteria in the procedure

- Establish and maintain procedures related to identifiable significant aspects of good and services used by the organization
- Communicate relevant procedures and requirements to suppliers and contractors

Documented operational control procedures and task instructions have long been part of an EMS—whether or not the EMS conforms to other parts of ISO 14001. What may be new to some organizations is the requirement to establish and maintain procedures related to the identifiable significant aspects of goods and services used by the organization and to communicate these relevant procedures and requirements to suppliers and contractors. Essentially, the organization must ensure that contractor activities—especially those performed on-site—do not conflict with the organization's environmental policy, objectives and targets, and overall EMS. To do this, the organization should consider methods of communicating relevant elements of the EMS to contractors. Example methods could include providing information about the EMS in the purchase contract, providing contractors with handbooks that detail the environmental policy and procedures that they are expected to adhere to while on-site, and performing informal reviews of the contractor's job site and communicating any nonconformances to the procedures.

EMERGENCY PREPAREDNESS AND RESPONSE (SECTION 4.4.7)

This part of ISO 14001 requires an organization to establish and maintain procedures to (1) identify potential for and respond to accidents and emergency situations and (2) prevent and mitigate the environmental impacts that may be associated with them. In addition, the organization must review and revise the emergency preparedness and response procedures where necessary, in particular after the occurrence of accidents or emergency situations. Finally, the standard requires the organization to periodically test these procedures where practicable.

CHECKING AND CORRECTIVE ACTION (SECTION 4.5)

To ensure the EMS is implemented as planned, objective and unbiased checking of EMS elements is needed. The crafters of the ISO 14001 standard provided proper emphasis on this within the numerous requirements of this section.

MONITORING AND MEASUREMENT (SECTION 4.5.1)

This element of the ISO 14001 standard requires that the organization establish and implement procedures to monitor and measure, on a regular basis, key characteristics of its operations and activities that could have a significant impact on the environment. When conforming to this requirement, the organization must record information to track performance, relevant operational control, and conformance with the organization's environmental objectives and targets. Although the organization must ensure that its environmental objectives and targets are tracked, this requirement goes beyond that by requiring monitoring and measuring activities for all key characteristics.

In addition to monitoring and measuring key characteristics, this element of the standard requires monitoring equipment to be calibrated per the organization's procedures and records of this activity to be maintained. The standard leaves it up to the organization to determine which pieces of monitoring equipment are specified in the procedures. Certainly, all monitoring equipment used for compliance to regulations and operational control should be included.

The final requirement defined in this section of the standard specifies that the organization must put in place a documented procedure for periodically evaluating compliance with relevant environmental legislation and regulations. Unlike the EMS audit, which is described fully within this book, this requirement does not necessarily require an audit, per se, and the evaluation does not have to be performed by an independent person.

NONCONFORMANCE AND CORRECTIVE AND PREVENTIVE ACTION (SECTION 4.5.2)

ISO 14001 requires that the organization establish and maintain procedures for defining responsibility and authority for handling and investigating nonconformances, for taking action to mitigate any environmental impacts, and for initiating and completing corrective and preventive action. The corrective and/or preventive action taken to eliminate the cause of the nonconformance should be appropriate to the magnitude of the problems encountered. Finally, the standard requires the organization to make changes to procedures, as necessary, as a result of corrective and preventive action.

RECORDS (SECTION 4.5.3)

Records are different from documents. Documents include procedures, instructions, manuals, and other forms of documentation that are used to manage the EMS. Records, on the other hand, are evidence that something has been accomplished (i.e., inspections, equipment calibration, and training.) ISO 14001 defines management of both documents and records, with control of documents being defined in Section 4.4.5 (see Chapter 1) and maintenance of records being defined in this section. In addition, the standard requires the organization to identify core elements of the EMS and the documents that interact with them; these requirements are outlined in Section 4.4.4—EMS Documentation.

Like many other sections of the standard, this section requires the organization to establish and maintain procedures—this time for the identification, maintenance, and disposition of records. The standard delineates that the organization must include training records and the results of the EMS audit and management reviews. Actually, the organization's environmental records typically will encompass much more than these. Examples of types of environmental records for which the above-mentioned procedures might apply are presented in Figure 1-2.

Examples of Environmental Records

✓ Applicable governmental regulations
✓ List of significant environmental aspects
✓ Environmental permits
✓ Training records
✓ Process hazard assessments
✓ Emissions modeling records
✓ Information on product attributes
✓ External environmental reports
✓ Inspection, maintenance, and calibration records
✓ Records of contractor activities on premises
✓ Incident and corrective action reports
✓ Records of testing of emergency procedures
✓ Records of compliance and EMS audit results
✓ Records of management review
✓ Reports to legal authorities
✓ Completed forms or checklists
✓ Letters or memos
✓ Organization chart

FIGURE 1-2.   Examples of environmental records.

Further, this section requires that environmental records be

- Legible
- Identifiable and traceable to the activity, product, or service involved
- Easily retrievable
- Protected against damage, deterioration, or loss
- Retained per established and recorded retention times

This requirement should pose no real burden on the mature EMS, although procedures for managing records may need to be expanded.

ENVIRONMENTAL MANAGEMENT SYSTEM AUDIT (SECTION 4.5.4)

EMS audits are required by ISO 14001 to assess whether or not the EMS conforms to planned arrangements for the EMS—including conformance to ISO 14001—and has been properly implemented and maintained. In specific, Chapters 2 and 5 of this book address specifics about EMS audits.

MANAGEMENT REVIEW (SECTION 4.6)

The final element of the ISO 14001 standard is that of management review. This section requires top management of the organization to review the EMS at specified intervals to ensure its continuing suitability, adequacy, and effectiveness. The standard further specifies that the management review process should ensure that necessary information is collected to allow management to carry out the evaluation. This information could include such things as environmental progress toward achieving objectives and targets; any accidents or incidents that had an adverse impact on the environment and corrective and preventive action; concerns of interested parties; results of EMS audits; and changes in activities, products, or services that might require changes in the EMS for it to

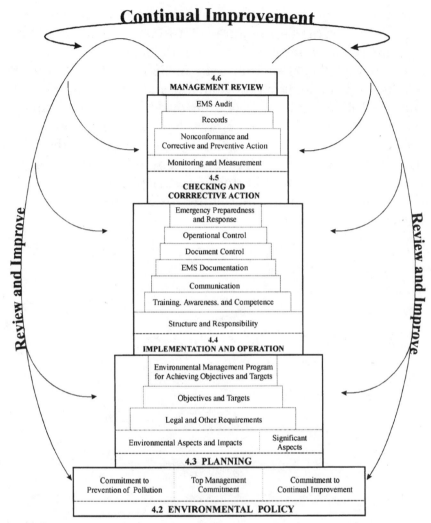

FIGURE 1-3.   Pictorial view of ISO 14001 framework.

remain suitable. In addition to reviewing information about the EMS, the standard requires that top management address the possible need for changes to policy, objectives and targets, and other elements of the EMS in light of the EMS audit results, any changing circumstances, and the commitment to continual improvement. It is expected that after the management review, the EMS will be revised to reflect the outcome of the review process. In particular, it is expected that the EMS itself will continually improve, thereby enhancing improvement of environmental performance. A pictorial view of the continual improvements that entire EMS framework fosters is presented in Figure 1-3.

ENVIRONMENTAL MANAGEMENT SYSTEM AUDITS

The international standard, ISO 14001: Environmental Management Systems—Specification with Guidance for Use, is built around the model of "environmental policy, planning, implementation and operation, checking and corrective action, and management review," with the ultimate intent of continual improvement (See Figure 1-1 in Chapter 1.) The importance of the "checking and corrective action" part of this model cannot be overstated, because what gets measured and checked (i.e., audited) typically gets appropriate management focus, and what does not get measured and checked easily gets forgotten. Thus, to implement the ISO 14001 standard effectively, it is necessary to understand fully the requirements for conducting the environmental management system (EMS) audit. This chapter provides this information and also discusses the relationship between EMS audits and environmental compliance audits.

THE AUDIT PROCESS

There have been many experts and standards organizations that have defined the term *audit*—whether it be quality audit or other types of audit—as follows:

- J. M. Juran defines a quality audit as "an independent review conducted to compare some aspect of quality performance with a standard for that performance" (Juran 1979).

- ISO 10011-1 standard defines a quality audit as "a systematic and independent examination to determine whether quality activities and related results comply with planned arrangements and whether these arrangements are implemented effectively and are suitable to achieve objectives" (ISO 1990).

- The Council of the European Communities in its Eco-Management and Audit Scheme (EMAS) has defined an environmental audit as "a management tool comprising a systematic, documented, periodic and objective evaluation of the performance of the organization, management system, and processes designed to protect the environment with the aim of: (a) facilitating management control of practices which may have impact on the environments; (b) assessing compliance with company environmental policies" (EMAS 1993).

- *Webster's Unabridged Dictionary* defines audit as "a methodical examination and review of a situation or condition concluding with a detailed report of findings" (Webster 1986).

These definitions provide some common threads that include what we consider to be the most important elements of an audit—systematic, independent (i.e., objective and unbiased), and a determination of whether planned arrangements are adhered to. Thus, the definition of an EMS audit can be summarized as a systematic, objective, and unbiased verification process to determine whether an organization conforms to planned arrangements for environmental management.

An EMS audit examines the organization's environmental management system with respect to the requirements of ISO 14001 and/or other self-imposed requirements. The purpose of the audit should determine if the EMS

- Meets the intent of ISO 14001
- Conforms to planned arrangements

- Is properly implemented
- Is properly maintained

What the EMS audit does not do is determine if the EMS is suitable, adequate, and effective. This is the responsibility of top management and is based on EMS audit results and other factors such as changing circumstances and the commitment to continual improvement.

The EMS audit results are used for corrective and preventive action and may provide an opportunity for continual improvement of the process. The audit process relies on the collection of objective evidence to prove conformance to ISO 14001 and planned arrangements. (Note: Objective evidence is the "proof" that the organization is doing what it says it is doing.) These planned arrangements can include compliance to legal/regulatory requirements, voluntary conformance to environmental standards such as ISO 14001, and/or self-defined environmental programs, procedures, instructions, and practices. Objective evidence can be gathered from many sources including documented procedures and programs, interviews with personnel who carry out the procedures and programs, environmental records, measurement and calibration data, permit applications, and other sources of information/evidence.

Persons from within an organization can conduct an audit of their own system provided that they did not establish or implement the elements of the system being audited, (i.e., they are objective and unbiased). This type of audit is a *first-party audit* and is often used for internal EMS audits required by the ISO 14001 standard. Another type of audit—a *second-party audit*—is typically a customer audit of a supplier or contractor to determine if the supplier or contractor conforms to prespecified contractual arrangements. A *third-party audit* is an audit conducted by an agency or independent registrar. This latter type of audit is typically not conducted as part of the internal EMS audit but rather as a compliance or registration or surveillance audit. (Note: Environmental compliance audits are discussed in greater detail later in this chapter.)

What is not acceptable when conducting an EMS internal audit is a *self-assessment* of one's own environmental processes. A self-assessment is an evaluation made by persons who are integrally involved with elements of the EMS (i.e., they have responsibility to establish, implement, or maintain the part of the EMS they are assessing). Thus, a self-assessment (or other type of audit) may be used to evaluate compliance with legal and regulatory requirements, as required in Section 4.5.1 of ISO 14001, but auditor objectivity is lacking, and this negates the use of this type of audit as a legitimate means of evaluating the EMS, as intended in Section 4.5.4 of ISO 14001.

ISO 14010: GUIDELINES FOR ENVIRONMENTAL AUDITING—GENERAL PRINCIPLES

A document specifying guidelines for environmental auditing has been developed by ISO. Although this document is not normative (that is, it is not required to be used during EMS audit), it does provide many useful principles for environmental auditing. Specifically, these include

- *Objective and scope*   The audit should be based on the objectives defined by the client.
- *Objectivity, independence, and competence*   The members of the audit team should be independent and should possess an appropriate combination of knowledge, skills, and experience to carry out the audit.
- *Due professional care*   The auditor should use care, diligence, skill, and judgment.
- *Systematic procedures*   Documented and well-defined methodologies should be used to enhance consistency and reliability.
- *Audit criteria, evidence, and findings*   Criteria should be agreed upon between the lead auditor and the client; appro-

priate information should be collected, analyzed, interpreted, and recorded; audit evidence should be of such quality and quantity that competent environmental auditors working independently of each other will reach similar audit findings.

- *Reliability of audit findings and conclusions* The desired level of confidence and reliability in audit findings and conclusions should be provided.
- *Audit report* A copy should be given to the client and should include items agreed upon by the client and the lead auditor.

PROTOCOL FOR EMS AUDITS

Elements of an audit protocol may vary slightly depending on the audit scope and whether or not it is a first-, second-, or third-party audit. However, typically these audits will include the following (1) preaudit activities such as a preaudit information gathering session or a preaudit visit, (2) planning of the audit, (3) official notification that the audit is to be conducted, (4) on-site activities, including an opening meeting, the audit, predetermined feedback sessions, and the closing meeting, and (5) an audit report.

Preaudit Activities

Preaudit activities allow the auditor(s) to get a sense of the organization's business, legal and regulatory requirements, and the general structure of the EMS. Documents that might be reviewed prior to beginning the audit include

- Organization mission statement
- Site map
- Environmental policy
- Legal and regulatory requirements, permits, orders of consent, certificates of operation, and other legal documents
- EMS manual (if the organization has elected to establish one)

- Operating procedures
- Organizational charts

If the auditor(s) deem it necessary, a previsit to the location may be needed to review the organization's activities, products, and services; this might be included as part of the preaudit activities. During this visit, the auditor(s) should confirm the audit scope and the time needed to conduct the audit.

Planning

Planning the audit is one of the key elements in the audit cycle. This element includes selecting the audit team, planning the audit agenda, providing training (if necessary) and information to the audit team, meeting with the audit team so that all members understand their roles, and developing and/or providing the audit methodology to the team. A sample audit agenda is presented in Figure 2-1. In addition, the auditee should agree to the agenda before the audit begins. Further, any special requirements need to be identified such as the use of safety glasses and/or shoes, personal protective equipment, special dietary needs, and other special requirements.

Notification

This element of protocol is typically done through written notice several weeks in advance, although some governmental agencies give notification that the audit is to be performed when they arrive on site. During the notification, audit team members are introduced, the date and time for the opening meeting is established, and the scope of the audit is provided.

Opening Meeting

During the opening meeting, auditors are formally introduced, auditee counterparts and/or guides are designated, and the audit agenda is presented. In addition, the lead auditor reemphasizes the audit scope and that the audit is only a sampling of overall activities, products, and services. Further, the lead

Tuesday, September 21, 1999

Audit Topic	Time	Location	XYZ Personnel
Opening Meeting	8:30-9:00 a.m.	Conference Room C	XYZ Top Management. EMS Management Rep., Environmental Staff, Legal Counsel, Key Area Managers/Reps.
Environmental Policy (4.2)	9:00 - 9:30	Conference Room C	XYZ Top Management, EMS Management Rep.,
- Aspects Identification (4.3.1) - Objectives & Targets (4.3.3) - Environmental Mgt. Program (4.3.4)	9:30-10:30 a.m.	Conference Room C	EMS Management Rep., Environmental Staff
Legal and Other Reqts	10:30 - 11:00 a.m.	Law Library	Env. Attorney
Structure & Responsibility (4.4.1)	11-11:30 a.m.	Environmental Staff Conference Room	EMS Management Rep.
Break - Lunch	11:30-12:30 p.m.		
Monitoring and Measurement (4.5.1) Energy consumption	12:30 - 1 p.m.	Environmental Staff Conference Room	Environmental Staff -- Program engineer for energy
Monitoring and Measurement (4.5.1) Wastewater discharges	1-1:30 p.m.	Environmental Staff Conference Room	Environmental Staff -- Program engineer for wastewater
Monitoring and Measurement (4.5.1) Hazardous waste discharges	1:30-2 p.m.	Environmental Staff Conference Room	Environmental Staff -- Program engineer for chemical/waste management
Monitoring and Measurement (4.5.1) Chemical management/spills	2-2:30 p.m.	Environmental Staff Conference Room	Environmental Staff -- Program engineer for chemical/waste management
Monitoring and Measurement (4.5.1) Air emissions	2:30-3 p.m.	Environmental Staff Conference Room	Environmental Staff -- Program engineer for air emissions
Emergency Planning and Response (4.4.7)	3-3:30 p.m.	Environmental Staff Conference Room	Environmental Staff -- Program engineer for chemical/waste management, Security
Monitoring and Measurement (4.5.1) HAZMAT transport	3:30-4 p.m.	Shipping/receiving office area	Shipping Manager
Auditor work session	4-4:30 p.m.	Conference Room C	N/A
Feedback Meeting w/ Auditors	4:30-5 p.m.	Conference Room C	All who participated in audit activities

FIGURE 2-1. Sample EMS audit agenda.

Wednesday September 22, 1999

Audit Topic	Time	Location	XYZ Personnel
Meet w/ Auditors to Review Nonconformances	8:00-8:15 a.m.	Conference C	EMS Management Rep., Environmental Staff, Other Affected Persons
- EMS Audit (4.5.4) - Nonconformance and Corrective/Preventive Action (4.5.2)	8:15-9:00 a.m.	Environmental Staff Conference Room	Environmental Staff -- EMS Audit Coordinator
- Training, Awareness, and Competence (4.4.2) - Communication (4.4.3)	9:00-10:0	Conference Room C	EMS Management Rep., Training Coordinator; Communications Coordinator
Operational Control Procedures/Records (4.4.6/4.5.3) Manufacturing Area	10:00-11:00	Manufacturing Line	Manager and Personnel from Manufacturing
- EMS Documentation (4.4.4) - Document Control (4.4.5) - Records (4.5.3)	11:00 - 12:15	Environmental Staff Conference Room	EMS Management Rep., Environmental Staff
Break - Lunch	12:50-1:00 p.m.		
Management Review (4.6)	1:00 - 1:45 p.m.	Top Management's Conference Room	Top Management, EMS Management Rep.
Operational Control Procedures/Records (4.4.6/4.5.3) Chemical/Waste Center	1:45 - 3:00	Chemical/Waste Center	Manager and Personnel from Chemical/Waste Center
Auditor's Meeting	3-4 p.m.	Conference Room C	N/A
Closing Meeting - Audit Summary	4-5 p.m.	Conference Room C	All involved with audit process

Thursday, September 23, 1999

Audit Topic	Time	Location	XYZ Personnel
Meet w/ Auditors to Review Nonconformances	8:00-8:45 a.m.	Conference C	EMS Management Rep., Environmental Staff, Other Affected Persons
Operational Control Procedures/Records (4.4.6/4.5.3) Utility Plants	8:45 - 12:00	Utility Plants A, B, and C	Manager and Personnel from the Utility Plants A, B, and C
Break - Lunch	12:00-1:00 p.m.		
Operational Control Procedures/Records (4.4.6/4.5.3) Contractors	1:00 - 3:00 p.m.	Contractors on Premises Areas: Cafeteria, graphics and copying centers, janitorial services	Appropriate Contractor Managers and Personnel
Auditor's Meeting	3-4 p.m.	Conference Room C	N/A
Closing Meeting - Audit Summary	4-5 p.m.	Conference Room C	All involved with audit process - management

FIGURE 2-1.   (*Continued*)

auditor sets forth the ground rules for the audit during this meeting and sets a tentative date and time for the closing meeting. Debriefing meetings about ongoing findings are also scheduled at this time. It is typical for the lead auditor to maintain a list of those who attended this meeting.

CONDUCTING THE AUDIT

Detailed information about conducting the audit to validate conformance to the EMS is presented in Chapter 5. Also included is an audit methodology, example audit questions, and objective evidence that might be used to perform the audit validation tests.

CLOSING MEETING

At the closing meeting the EMS audit results are reported to appropriate personnel, including management. Typically, it is at this meeting that management and/or the management representative understands and agrees to the findings and commits to corrective and/or preventive action to eliminate any nonconformances. Additionally, the lead auditor identifies the approximate time in which the audit report will be completed.

AUDIT REPORT

The audit team leader should gather information from all team members and prepare an audit report. An example audit report is presented in Appendix A. This example represents a very detailed EMS audit report and contains audit scope, reference to audit methodology, auditor qualifications, agenda, nonconformance/observation forms—including a summary form—and auditor notes describing areas/departments tested. This type of report might be written during a first- or second-party audit. During a third-party audit, the agenda and nonconformance/observation forms are typically left with the organization at the time of the audit. The written report that follows is a synopsis of the audit findings, including observations for continual improvement. Auditor notes typically are not included.

REQUIREMENTS OF ISO 14001 WITH RESPECT TO EMS AUDITS

Requirements for establishing and implementing an audit program and procedures and for conducting an EMS audit are found in Section 4.5.4 of ISO 14001. They include

- The requirement for the organization to establish and maintain (a) program(s) and procedures for periodic EMS audits. The audit program and schedule must be based on the environmental importance of the activity concerned, and the audit procedures must cover the audit scope, frequency, methodologies, and responsibilities and requirements for conducting audits and reporting results.
- The requirement for the EMS audits to determine whether or not the EMS conforms to planned arrangements for environmental management including the requirements of ISO 14001. In addition, EMS audits must determine whether or not the EMS has been properly implemented and maintained.
- The requirement for providing information on audit results to management. In particular, these results must be reported to top management during the management review defined in Section 4.6 of ISO 14001.

There are no requirements in the ISO 14001 standard for how to conduct an EMS audit; however, most auditors (especially those who are professional auditors) will agree that following a standard protocol—such as that described in the above section—and having a defined methodology—which is required by the standard—can make the job of auditing the EMS consistent and efficient. A discussion related to the EMS audit program and procedures is presented in Chapter 4. Information about developing an EMS audit methodology and conducting the EMS audit is provided in Chapter 5.

NONCONFORMANCE AND CORRECTIVE AND PREVENTIVE ACTION

Part of the audit process includes the auditors' identification of nonconformances to ISO 14001 and/or requirements of the EMS. The auditee must have a procedure for handling and investigating these and other identified nonconformances, for taking action to mitigate any impacts, and for initiating and completing corrective and preventive action. A sample procedure for this element is presented in Figure 2-2. Examples of types of nonconformances that might be identified within the EMS are presented in Figure 2-3. Handling and investigating nonconformances—including environmental incidents—might be achieved with a simple nonconformance tracking log, an example of which is shown in Figure 2-4. As can be seen from the figure, nonconformance, root cause, corrective/preventive actions, responsible person, and target date for closure are included. The log should be updated on a regular basis (i.e., weekly or biweekly) and can provide for easy identification of recurring problems.

THE RELATIONSHIP BETWEEN EMS AUDITS AND COMPLIANCE AUDITS

There has been considerable discussion about the relationship between EMS audits and legal and regulatory compliance audits. In particular, governmental bodies within the United States have argued that EMS audits do not adequately address environmental compliance to legal and regulatory requirements. There are many who view compliance audits as a completely different type of audit from that of an EMS audit.

In fact, this is not the case at all. The major difference between the two audits is the approach that the auditor takes. Typically, a compliance auditor verifies environmental results through "end of pipe" testing. Environmental data, permits, reports, shipping papers, and other environmental documents are reviewed to ensure that the organization has stayed within

Procedure Name: Nonconformance and Corrective and Preventive Action
Document Control Number: EMS 4.5.2
Document Owner/Approver: Mary Smith, Environmental Manager
Date: 04/12/99

Introduction.

XYZ is committed to taking corrective and preventive action to mitigate nonconformances identified within the environmental management system (EMS). XYZ has taken such measures over the years when emergency situations have occurred, and is expanding this practice to include nonconformance to all elements of the EMS. Corrective and preventive action will be appropriate to the magnitude of the problem (potential or actual) identified and will be in line with the environmental policy and environmental impact of the nonconformance.

Requirements and Responsibilities.

Requirements and responsibilities for nonconformance and corrective and preventive action at the Corporate and plant site levels are as follows:

The Environmental Manager is responsible for handling any nonconformance to the EMS identified during the EMS audit process or through other means. This person will investigate the nonconformance using root cause analysis and will develop a plan for corrective and preventive action. Should human or financial resources be needed to initiate the plan, these must be authorized by the General Manager of XYZ Once the corrective and preventive action plan is initiated, the Environmental Manager will document and track all such actions to closure.

FIGURE 2-2.   XYZ's procedure for managing nonconformance and corrective and preventive action.

**Types of Nonconformances that
Might be Identified During an EMS Audit**

Environmental Policy
✓ Activities or operations do not support the environmental policy
✓ Environmental policy does not contain commitment to continual
   improvement of the EMS

Planning
✓ Significant environmental impacts have not been defined
✓ Employees whose job functions could affect legal requirements do
   not have access to these requirements
✓ Views of interested parties were not considered when setting objectives
   and targets
✓ Environmental management program does not specify responsible
   person

Implementation and Operation
✓ Defined roles, responsibilities, and authorities have not been
   communicated to relevant employees
✓ Training plan is not being followed
✓ Procedures for receiving, documenting, and responding to relevant
   external communications are not established
✓ Direction to those documents that interact with the core elements of
   the EMS is not provided
✓ Obsolete documents were not removed promptly from points of use
✓ Contractors were not aware of organization's procedures that related
   to the EMS
✓ Emergency plan was not tested as prescribed in procedures
✓ Equipment calibration schedule was not followed

Checking and Corrective Action
✓ Responsibility for handling nonconformances to the EMS was not
   defined
✓ Records were not easily retrievable
✓ The EMS audit schedule was not defined

Management Review
✓ The management review was not documented

FIGURE 2-3.   Examples of types of nonconformances that
might be identified during an EMS audit.

NONCONFORMANCE TRACKING LOG
FOR EMS AUDIT CONDUCTED 07/13-07/15, 1999
Last update 08/06/99

Nonconformance	Root Cause	Corrective/Preventive Action	Person Responsible	Target Date(s)
Environmental policy does not contain commitment to continual improvement.	New requirement as a result of implementing ISO 14001.	Update environmental policy to add commitment to continual improvement.	Env. Mgr. to draft update; top mangmt. to issue new policy	08/15/99; 10/15/99
One employee at the Wastewater Treatment Facility did not know permit requirements or operating limits.	New employee training did not address permit requirements and operating parameters.	Add permits requirements and operating parameters on new employee training checklist; train new employee on this information; post this information in control room.	Department Mgr. of Wastewater Treatment Facility	08/01/99 (Closed)
All roles and responsibilities within the EMS are not defined – i.e., the roles of manager/ employees of wet process line were not defined in terms of achieving objectives and targets.	Process for identifying roles and responsibilities for implementation of the EMS did not include the entire organization, only the environmental staff and management representative.	Add roles of manager/ employees of wet processing line in terms of achieving objectives and targets to EMS manual; evaluate all departments and contractors activities to determine interaction with the EMS and define roles and responsibilities, as appropriate.	EMS Coordinator	10/15/99

FIGURE 2-4.   Sample nonconformance tracking log.

NONCONFORMANCE TRACKING LOG
FOR EMS AUDIT CONDUCTED 07/13-07/15, 1999
Last update 08/06/99

Nonconformance	Root Cause	Corrective/Preventive Action	Person Responsible	Target Date(s)
Retention times for all environmental records had not been identified.	EMS training did not adequately address the requirements for identifying environmental records.	Improve training to adequately address the requirements for identifying environmental records.	EMS Coordinator	10/30/99
Calibration records for on-line pH meters were not available.	Personnel responsible for calibrating pH meters were not aware calibration records should be maintained as environmental records.	Calibration records for pH meters will be maintained; department manager will identify if other calibration records must be maintained; employee training on environmental records management will be enhanced.	Department Mgr. of Wastewater Treatment Facility	09/01/99

FIGURE 2-4. (*Continued*)

NONCONFORMANCE TRACKING LOG
FOR EMS AUDIT CONDUCTED 07/13-07/15, 1999
Last update 08/06/99

Nonconformance	Root Cause	Corrective/Preventive Action	Person Responsible	Target Date(s)
Employees on the manufacturing line were unaware of the environmental policy as it relates to their job responsibilities; this was inconsistent with XYZ's procedure.	Managers did not deploy EMS training as required..	Manufacturing line employees will be given EMS awareness training; all department mangers have been asked to verify that employees in their department are aware of the EMS and their role in the EMS; additional requirements will be added to the organization's EMS audit methodology to verify EMS training has been completed.	Manager of Human Resources	08/15/99
Several operating procedures at the chemical distribution center were not approved as required by XYZ's procedure.	Employees were unaware that the documents had to be approved; employee training for document control was not adequate.	Procedures at the chemical distribution center have been reviewed and approved; all department managers have been asked to ensure that environmental documents and procedures have been approved; employee training on document control will be improved.	EMS management representative	08/31/99

FIGURE 2-4.   (*Continued*)

legal and regulatory bounds. Essentially, the focus is an after-the-fact review of required implementation. An EMS auditor, on the other hand, reviews conformance to the whole system, including policy, planning, implementation, operation, checking, and management review. (Note: In many countries or regional geographies, the terms *conformance* and *compliance* are used interchangeably. However, ISO 14001 explicitly uses the term *compliance* with respect to legal and regulatory requirements and *conformance* with respect to the EMS.) This latter approach focuses on proactive environmental management from top to bottom and within the system.

Legal and regulatory compliance is an integral part of an overall EMS and, likewise, auditing compliance is an integral part of the EMS audit. Thus, simply put, the relationship between an EMS audit and a legal and regulatory compliance audit is that the compliance audit is a subset of the EMS audit. How much focus legal and regulatory compliance gets during the EMS audit can be set forth in the audit scope. Certainly, there are many opportunities to include audit compliance elements within the EMS audit. Examples of evidence of legal and regulatory compliance that might be reviewed within the structure of the EMS audit are presented in Table 2-1.

TABLE 2-1. Examples of Evidence of Legal//Regulatory Compliance that Might be Reviewed within the Structure of the EMS Audit

Commitment to Comply with Legal Requirements (Section 4.2)

- Identify any notices of violation, fines, or enforcement orders.
- Review how the organization evaluates whether or not it is meeting legal and regulatory requirements.
- Review the last governmental agency compliance audit for air emissions, hazardous waste, wastewater discharge, etc. (Note: If there are compliance problems noted, review a sample of corrective actions to verify the problem was mitigated and corrected.)

TABLE 2-1. Examples of Evidence of Legal//Regulatory Compliance that Might be Reviewed within the Structure of the EMS Audit (*Continued*)

Identification of Legal Requirements (Section 4.3.2)

- Review a sample environmental permit for air emissions, hazardous waste, wastewater discharges, and so forth, to test that they are up to date.
- Review a sample of environmental permit applications to test for accuracy.
- Test a sample of permit requirements to verify that they are being met.
- Review a sample of legally required reports to verify that they are accurate and filed on time.
- Review how the organization identifies legally required training.

Setting Objectives and Targets (Section 4.3.3)

- Select a sample of objectives and/or targets that are legal or regulatory requirements and verify that they are being met. (Note: This could include objectives and targets such as waste minimization objectives and targets, air emissions reduction objectives and targets, best management objectives for stormwater runoff, and recycling objectives and targets.

Defining Responsibility (Section 4.4.1)

- Review a sample of environmental permit applications to test that they were signed by the appropriate person, as required by regulation.

Ensuring Proper Training and Competence (Section 4.4.2)

- Test a sample of employees and review training records to ensure that they have received legally required training.
- Select a sample of employees and verify that competence was assessed by the organization per legal requirements.
- Select a sample of training programs to determine if they conform to the legally required training content.

TABLE 2-1. Examples of Evidence of Legal//Regulatory Compliance that Might be Reviewed within the Structure of the EMS Audit (*Continued*)

Providing External Communication (Section 4.4.3)

- Review a sample of legally required external communications—such as public notices, external incident reporting, and other required communications—to verify that they were provided.

Ensuring Proper Documentation (Section 4.4.4)

- Review a sample of legally required documents to verify that they are available.

Ensuring Operational Control (Section 4.4.6)

- Review a sample of shipping papers to verify that they are accurate.
- Review a sample of legally required plans (such as spill prevention plans, stormwater runoff plans, waste management plans, etc.) to verify that they are being followed.
- Review a sample of employee activities with respect to chemical and waste handling to verify that they meet legal and regulatory requirements.
- Review a sample of chemical and waste labels to verify that they meet legal and regulatory requirements.
- Review a sample of legally required reports to verify they are accurate and filed on time.

Providing for Emergency Preparedness and Response (Section 4.4.7)

- Review the legally required emergency plan(s).
- Review a sample of legally required emergency tests.
- Review a sample of incidents and response to verify that the response conforms to emergency plans.
- Review a sample of external reports of incident to verify that they are accurate and filed on time.

Table **2-1.** Examples of Evidence of Legal//Regulatory Compliance that Might be Reviewed within the Structure of the EMS Audit (*Continued*)

Obtaining Monitoring and Measurement Data and Evaluating Compliance to Legal/Regulatory Requirements (Section 4.5.1)

- Review a sample of legally required wastewater discharge reports, air emissions reports, waste reports, and so forth, to verify that the data is accurate and that required parameters have been monitored and/or measured.

- Review a sample of legally required wastewater discharge reports, air emissions reports, waste reports, and so forth, to verify that permit requirements are met.

- If there are requirements for a governmental agency to certify laboratories, review certification records.

- Review a sample of laboratory data—including quality control, quality assurance, and holding times—to ensure that laboratory data provided to meet legal and regulatory requirements is valid.

- Review a sample of equipment calibration records for laboratory data to verify the equipment used for provide data for legal and regulatory requirements is calibrated per procedure.

- Review a sample of equipment maintenance records to verify that the sampling equipment is properly maintained.

- Review how the organization evaluates compliance with legal and regulatory requirements.

- Review a sample of the compliance evaluations and verify that any noncompliance situations were corrected and that notification to governmental authorities was made, if required.

Nonconformance and Corrective and Preventive Action (Section 4.5.2)

- Review action(s) taken to correct noncompliance issues.

Records (Section 4.5.3)

- Review a sample of legally required environmental records—including inspection records, training records, laboratory reports, and reports to governmental agencies—to verify that they are available, as required, and complete.

REFERENCES

Council of the European Communities, *Eco-Management and Audit Scheme,* 1993.

International Organization for Standardization, ISO 10011-1, Geneva, 1990.

Juran, J. M. (ed.), *Quality Control Handbook,* McGraw-Hill, New York, 1979.

Grove, P. B., *Webster's Unabridged Dictionary,* Merriam-Webster, Springfield, Mass., 1986.

AUDITOR AND AUDITEE RESPONSIBILITIES DURING AN EMS AUDIT

Auditors and the auditee both have roles and responsibilities within the audit process to make the environmental management system (EMS) audit effective and efficient. Below are some guidelines about auditor qualifications, auditor responsibilities, auditee responsibilities, and some "do's and don'ts" for both auditor and auditee.

AUDITOR QUALIFICATIONS

It cannot be overstated that the internal EMS audit is a critical element of the overall management system because results of this audit are used by top management to determine if the EMS is suitable, adequate, and effective. Thus, it is important that those conducting the EMS audit(s) are qualified to do so. Because the effectiveness (or ineffectiveness) of the overall EMS has great potential to significantly affect the environment, it follows that EMS auditor training needs should be identified and levels of competence defined, as required in Section 4.4.2 of the standard under "training, awareness and competence." In addition, auditor qualifications should be defined in the EMS audit procedure(s) and program(s), under Section 4.5.4.

Things to consider when defining auditor qualifications include

- Size and complexity of the organization
- Business mission
- Available internal resources for conducting the EMS audit
- Availability and cost of training classes
- Potential for hiring a consultant to conduct the audit
- Registration intentions

Examples of skills and training frequently identified as requirements for auditors include

- Knowledge of the organization's EMS and the requirements of ISO 14001
- Knowledge of the EMS audit methodology
- Environmental experience
- Audit experience and/or training
- Knowledge of applicable legal requirements
- Interpersonal skills
- Writing skills
- Communication skills

Although not specified in the standard, the person charged with conducting internal EMS audits should receive some form of specialized management system audit training. This specialized training is important to ensure that the auditor is able to perform the EMS audit effectively and efficiently. Depending on the complexity of the system, this training can range from a 1-day overview to a 5-day lead auditor training course, and more. As a minimum, those conducting the internal EMS audit should have some basic understanding and training on the following topics:

- Basics of ISO 14001 (see Appendix B for outline of this)
- Audit planning
- Preparing an agenda
- Developing an audit methodology
- Conducting an opening meeting
- Conducting the audit and interviewing personnel
- Writing and communicating nonconformance reports
- Conducting a closing meeting
- Writing a final report
- Follow-up and closure of nonconformance reports

Example EMS auditor training overheads are presented in Appendix B.

AUDITOR RESPONSIBILITIES

The audit team has numerous responsibilities when conducting an EMS audit. First, the team must follow a standard protocol, an outline of which is presented in Chapter 2. The following paragraphs present the responsibilities of the lead auditor, auditor, and—if included as part of the audit team— technical expert.

Lead Auditor

The lead auditor is responsible for coordinating all EMS audit activities. Examples of lead auditor responsibilities include

- Select audit team
- Prepare audit protocol and methodology
- Serve as liaison between auditee and audit team

- Verify audit scope with auditee and ensure that all elements of this scope are audited
- Serve as focal point for handling technical issues and/or disputes
- Write audit report
- Close nonconformance when evidence of closure is adequate

In addition to these responsibilities, the lead auditor also performs the role of auditor and has those responsibilities as well.

Auditor

Responsibilities of auditor include the following:

- Follow established audit protocol and methodology
- Verify that the EMS conforms to the intent and requirements of ISO 14001 and the organization's management system
- Write clear and succinct nonconformance findings and articulate such to the auditee

Technical Expert

Although not always required, a technical expert is included as part of the audit team if he or she is needed to impart specific knowledge about industry-specific, regulatory, and/or other EMS issues. The technical expert, in essence, plays the role of a "consultant" to the team and typically does not write nonconformance findings.

Auditor Do's and Don'ts

The audit team should be especially aware of the fact that many who are being audited look upon auditors as those who are trying to expose weakness in their work efforts. To overcome this stigma and to make the audit process effective and efficient, auditors should do the following:

- *Be prepared* Ensure that the audit team is qualified, have a defined audit protocol and methodology, and review the

EMS manual (if the organization has one) and key environmental procedures before beginning the audit.

- *Confine the audit to the agreed-upon scope* Don't follow the audit trail into areas that are outside of the audit scope.

- *Remain objective* Audit each location or area against the requirements of ISO 14001 and internal location requirements; don't compare one location or area to another.

- *Utilize optimal audit questioning techniques* Be a good listener, keep questions brief and to the point, ask open-ended questions (i.e., who, when, where, how, why, and what), and ensure that the questions gets answered.

- *Optimize personal attributes* Be professional, courteous, and friendly.

- *Ensure proper protocol* Keep to the audit agenda, perform proper roles (i.e., auditor, observer, or technical expert), and clearly articulate and document audit findings.

Auditors should *not* do the following

- *Bring in personal aspects* Don't be confrontational, don't react to others' personalities, don't argue, don't pinpoint blame for a nonconformance, and don't be arrogant.

- *Lead the auditee* Don't ask closed-ended questions (i.e., yes or no), don't consult, and don't refer to what you think is a "better" way of doing things.

AUDITEE RESPONSIBILITIES

Above, we described the qualifications and responsibilities of the auditor. The auditee, likewise, has responsibilities during the EMS audit. Typically, the auditee will appoint an audit coordinator, who will handle logistics, ensure that required information is gathered and provided, and ensure that feedback from the auditors is provided to the proper persons. Other persons who will likely be involved as auditees during the EMS audit include top management; the management representative; environmental, legal, and communications staff; full-time, part-time, and temporary employees of all job categories; and contractors.

RESPONSIBILITIES OF THE AUDIT COORDINATOR

Being the audit coordinator is a very important task, and the person selected for this job should be knowledgeable about the organization's business and EMS. In addition, this person should be able to work well with people, cope with changing plans, and coordinate multiple activities within the time frame of the audit. Essentially, this person is the "face" of the organization being audited, and it is always beneficial for the auditee to give a good impression to the auditor—no matter what type of audit is being conducted. Some specific tasks that the audit coordinator might be expected to perform are presented in Figure 3-1.

ROLES AND RESPONSIBILITIES OF TOP MANAGEMENT

Top management has unique responsibilities during the EMS audit because it is top management who defines the environmental policy and reviews the EMS for suitability, adequacy, and effectiveness. Top management should attend the opening and closing meetings and should be interviewed as part of the audit. Scheduling should allow enough time for the auditor to feel comfortable that top management is integrally involved with the EMS and has a rationale for determining suitability, adequacy, and effectiveness of the EMS.

ROLES AND RESPONSIBILITIES OF EMS MANAGEMENT REPRESENTATIVE

Roles and responsibilities of the EMS management representative are expressly called out in the ISO 14001 standard. Essentially, the EMS management representative is appointed by top management to ensure that the EMS is established, implemented, and maintained according to the ISO 14001 standard. In addition, this person is responsible for reporting on the performance of the EMS to top management for review and as a basis for improvement of it.

**Tasks that the Audit Coordinator
Might be Expected to Perform**

Prior to Audit

✓ Work with lead auditor to set dates and agenda for audit
✓ Provide lead auditor audit scope, mission, and other introductory
  information
✓ Schedule needed offices and conference rooms
✓ Provide needed equipment such as phones, FAX machines, and
  computer equipment
✓ Identify and provide safety equipment needed during the audit
  such as safety shoes, hard hats, safety glasses, etc.
✓ Identify any special needs of the auditors such as dietary
  needs/restrictions or medical conditions/needs
✓ Communicate agenda to affected managers and staff
✓ Identify the need for participation of second and/or third shift
  employees and managers and notify appropriate managers
✓ Ensure auditees understand the audit process

During Audit

✓ Ensure agenda is followed to the extent possible
✓ Keep all informed of changes to schedule
✓ Collect required documents for auditors, as necessary
✓ Provide guides for auditors
✓ Record followup items and ensure closure
✓ Schedule feedback meetings and communicate audit feedback to
  appropriate persons

After Audit

✓ Review audit report
✓ Brief management representative(s) of audit results
✓ Ensure corrective and preventive action process is in place
✓ Identify areas in EMS that could be improved upon

FIGURE 3-1. Tasks that the audit coordinator might be
expected to perform.

ROLES AND RESPONSIBILITIES OF ENVIRONMENTAL STAFF

The environmental staff members play a key role during the EMS audit, because they are the people who establish and implement the environmental management programs. In addition, they are typically the ones who are responsible for evaluation of compliance with legal and regulatory requirements. These people are expected to be totally familiar with the EMS—particularly with significant aspects, objectives and targets, and roles and responsibilities within the framework of the EMS. Often, persons from the environmental staff will serve as guides for auditors reviewing operations areas related to their environmental program (i.e., the person responsible for waste management will accompany the auditor to the waste handling area or areas).

ROLES AND RESPONSIBILITIES OF EMPLOYEES AND CONTRACTORS WHOSE JOBS HAVE THE POTENTIAL TO IMPACT THE ENVIRONMENT

These full-time, part-time, and contract employees should be aware of organization's environmental policy and how it relates to their jobs. If there are legal requirements that their actions can affect, they should understand these as well. In addition, they should understand the organization's objectives and targets that they can influence. And, finally, they should know and be able to explain what to do in an emergency situation and whom to contact if they have any questions regarding environmental issues.

GUIDES

Guides play an important role in the audit process. In addition to helping the auditor navigate to the appropriate department or area, they also have the responsibility to maintain the audit schedule. Guides help keep other auditees focused on the auditor's questions and offer assistance if there is confusion on the part of the auditor or auditee. Further, they have the responsibility to take notes, if necessary, when the confusion is not cleared up. Essentially, they are the objective "eyes and ears" of the EMS management representative.

Legal Counsel

The organization's internal or external legal department or support is expected to understand the environmental policy commitments and their support of them. In addition, they should understand legal requirements and provide details of how they make these accessible and/or communicate the relevant information to those who have a need to know.

Communications Manager

The communications manager typically has the responsibility to ensure that internal communication about the EMS—such as bulletin board announcements, newsletters, and so forth—are published as planned. In addition, this person will probably keep documentation of external communications (and responses or the decision not to respond) from relevant external parties.

Do's and Don'ts for Auditees

There are many things that the auditee can do to make the EMS audit proceed smoothly; likewise, there are some actions that the auditee should avoid during the audit. Actions of the auditor that are useful or beneficial include the following:

- *Be courteous* Remember the EMS auditor is only testing what the organization says it does. The auditor is not trying to highlight personnel ineffectiveness or present "out of control" situations to management. Further, if it is a third-party audit, the organization's management (or representative) invited the auditors, so discourteous behavior will be embarrassing to the entire organization.

- *Be punctual* Those being audited should not leave the auditors waiting because this is discourteous to the auditors and unfair to other people on the agenda. If there is occasion when there is a real business need for the auditee to delay or miss the scheduled meeting, that person should give as much notice as possible. If advance notice is not possible, someone should be available during the scheduled audit time to let the auditor (and guide) know that the meeting must be rescheduled.

- *Answer only the question that is asked*   The auditor typically will allow the auditee to "drift" to additional topics, which may lead the auditor to areas beyond the given scope. The auditee should be specific and answer the question that was asked. If not sure of the question, the auditee should ask the auditor to repeat it or rephrase it.

- *Answer questions with confidence and self assurance*   The auditee should be able to answer questions about their part of the EMS without having to look at a manual, procedure, record, and so forth. If the auditee is prepared, he or she should be able to respond with a "Yes, let me show you" or "Yes, here it is." These answers are always preferable to the auditee's giving the auditor a blank look or a response of "Wait here while I see if I can find it."

There may be times when the auditee is asked a question outside the realm of his or her experience. In these cases, the auditee should simply explain that he or she will check with the person responsible for that aspect of the EMS. In addition, as the auditors ask questions and verify answers, they are also watching body language, so the auditee should smile and exude confidence.

- *Discuss observations and nonconformances with auditor*   Ensure there is a clear understanding of any nonconformances or observations the auditor raises. In cases of uncertainty, the auditee should ask for clarification of specific observations and/or nonconformances.

- *Keep management informed*   As most professionals who have worked in industry have learned, management does not like surprises. Before the closing meeting management should be briefed about what they will hear.

- *Keep a sense of humor*   The organization is developing a partnership with the auditor(s). Everything is not going to go perfectly, so be able to laugh about auditee mistakes.

The auditor should not do the following:

- *Don't take things personally or be defensive*   The auditee typically knows his or her system very well, but how the EMS works may not be obvious to auditor or others. Auditors are trained to ask questions, to verify answers with objective evidence—such as documents or records—and to observe. They are also trained to ask open-ended questions, even if they know the answer. The auditor typically will start each question with how, what, when, where, why, or who. Once the answer is provided, the auditor will probably say "Show me".

- *Don't be argumentative*   Often, if the auditee thinks the current way of operating is the "right" way, he or she may argue trivial (or not so trivial) points with the auditor. This does nothing for the audit process or the EMS. The auditee should stay calm and try to understand the points that the auditor is making.

- *Don't question other auditees in front of the auditor or argue with others in the organization*   Invariably an auditee will answer a question that some other person within the organization would like to see answered differently. The temptation is for that person to further expand on the answer. This type of follow-up is acceptable if conducted out of the presence of the auditor but should be avoided when the auditor is present.

   In addition, there will be instances when auditees disagree on an answer, procedure, or other form of objective evidence. Again, agree to discuss these matters at a later time, and don't argue in front the auditor.

- *Don't answer for other people*   The auditee should restrict comments to his or her area of responsibility and expertise.

- *Don't criticize others within the organization or point out shortcomings of the EMS*   Some employees like to take the opportunity during an audit to vent frustration. The audit coordinator should "screen" those employees likely to be audited and should ensure that all employees understand that their role in the audit is to answer the questions and not make peripheral comments about inadequacies of management or the system.

- *Don't leave the auditor alone during the audit*   A guide should accompany the auditor at all times. This will ensure that the auditor does not go into a restricted or unsafe area. In addition, the guide can provide an additional service of listening to all conversations between the auditor and auditee in case a question about what was said arises later.

- *Don't bring the auditor to an area unannounced*   Nobody likes surprises, especially during an audit. If for some reason the auditor has to talk to a person that is not on the original agenda (which is actually quite common), ensure that the person is contacted so that time can be scheduled. In addition, it is always a good idea to inform him or her about what will be asked.

- *Don't accept a nonconformance if it is vague or incorrect*   If the nonconformance is not clearly understood or (worse yet) inaccurate, discuss this with the auditor. Otherwise, it will be difficult to implement appropriate corrective and preventive action. Request that the auditor be as specific as possible when writing a nonconformance.

GUIDANCE FOR PLANNING AND CONDUCTING AN EMS AUDIT

INTERNAL EMS AUDIT PROGRAM AND PROCEDURES

EMS AUDIT PROGRAM AND PROCEDURES

Section 4.5.4 of ISO 14001, which specifically addresses the environmental management system (EMS) audit, includes the requirement for the organization to establish and maintain a program(s) and procedure(s) for these EMS audits to be carried out periodically. Annex A of the standard—which provides guidance—suggests that the program and procedure(s) should cover the following:

- The activities and areas that will be reviewed or considered for review during the audits
- The frequency of the audits
- The responsibilities associated with managing and conducting the audits
- The communication of audit results
- Auditor qualifications, training, and competence
- How the audit will be conducted (audit methodology)

A sample audit procedure and program are presented in Figures 4-1 and 4-2, respectively.

Procedure Name: Environmental Management System (EMS) Audit
Document Control Number: EMS 4.5.4A
Document Owner/Approver: Mary Smith, Environmental Manager
Date: 05/18/99

XYZ is committed to assuring that its EMS functions properly. In order to do this, the EMS must be audited by an auditor or team of auditors who are objective and unbiased. Requirements and responsibilities for auditing the EMS at XYZ Company are as follows:

a. The Environmental Manager is responsible for initiating the EMS audit process at XYZ Company. The audit scope, expected frequency, and methodology is determined by this person and is established in the Environmental Management System (EMS) Audit Program, which is defined in the document EMS 4.5.4B. It is the responsibility of the Environmental Manager to be the focal point of interface between XYZ Company and the audit team.

b. The EMS audit team will consist of: XYZ Legal Counsel (lead auditor); a member(s) from the quality auditing team who has had ISO 14001 auditor training; and a member(s) who has three years environmental experience and who is outside the area being audited. It is the responsibility of the lead auditor to conduct the audit in accordance with the XYZ audit methodology and to summarize the findings in an audit report and present it to the President of XYZ Company.

c. Results of the EMS audit will be documented in the XYZ standard audit report format.

d. EMS audit reports will be kept for five (5) years and disposed of as confidential (recycled) waste paper.

FIGURE 4-1.   XYZ company's procedure for auditing the EMS.

Environmental Management System (EMS) Audit Program
Document Control Number: EMS 4.5.4B
Document Owner/Approver: Mary Smith, Environmental Manager
Date: 05/18/99

EMS Audit Scope.

The EMS audit program covers XYZ Companies activities products, and services. All elements of the EMS will be audited, with special emphasis placed on the following:

(1) Awareness and support of the environmental policy and communication of it to employees and contractors
(2) Environmental aspects and significant environmental aspects identification
(3) Understanding of objectives and targets at all relevant levels
(4) Training and awareness methods to ensure that employees understand how their job functions impact the environment and the overall EMS, including the environmental policy and objectives and targets
(5) Communication about the EMS at all relevant levels
(6) Compliance and contractor audits
(7) Adherence to operational procedures
(8) Handling of nonconformance and corrective and preventive action
(9) Management review

EMS Audit Frequency.

A complete EMS audit will be conducted at least annually. Specific elements will be tested more frequently as audit findings warrant.

FIGURE 4-2.   XYZ company's EMS audit program.

Environmental Management System (EMS) Audit Program *Cont.*
Document Control Number: EMS 4.5.4B

Results of Audits.

EMS audit results at XYZ Company will be reviewed by the President of XYZ Company to determine if the scope and/or frequency of the audits needs to be changed.

Responsibilities and Requirements for Auditors.

The XYZ Legal Counsel will act as the lead auditor on all EMS audits. This person will follow typical audit protocol. Other audit team members consist of quality auditors who have had ISO 14001 auditor training and a member who has at least three years of environmental experience.

Audit Methodology.

The XYZ audit methodology is defined in *ISO 14001 Implementation Manual*, "Appendix B," McGraw-Hill, 1998.

FIGURE 4-2. (*Continued*)

EMS AUDIT PLANNING AND SCHEDULING

EMS audits are a vital component of the environmental management process because these audits test whether or not the EMS is working as planned. It is important, therefore, to have a comprehensive plan and well-thought-out schedule for EMS auditing.

When audits are planned and scheduled with full participation and knowledge of the departments, functions, and personnel being audited, they will proceed efficiently and openly. Because audits typically affect workload and resources, proper planning and scheduling is critical to minimize disruption of the business. Some do's and don'ts when planning and scheduling the EMS audit are presented in Figure 4-3.

When preparing the audit plan and schedule, the following should be considered:

- The environmental importance of the activity, including the potential environmental impact
- Results of previous audits
- Sampling size of personnel and documents to be audited
- Schedules of other audits that are being conducted in the area, such as quality audits and health and safety audits

The audit plan and schedule may be developed to allow the entire system to be tested during a single event, or the schedule may select different elements of the standard and different departments and/or functions at different times. Examples of audit plans and schedules are presented in Figures 4-4 through 4-6, with the first figure representing a complete EMS audit conducted during one session and the second one representing a series of partial audits of the elements of the EMS conducted over time.

Planning and Scheduling an EMS Audit

Do

✓ Plan for a complete EMS audit to be conducted during the time frame established in the EMS audit procedure
✓ Coordinate the EMS audit schedule with other planned audits such as quality and health and safety to minimize impacts to the business
✓ Ensure auditors conducting the audit are qualified
✓ Communicate the audit schedule in advance
✓ Remind auditees that the EMS audit is a friendly audit and that the purpose of the audit is to improve the EMS, not assign blame for nonconformances

Don't

✓ Be inflexible when developing the EMS audit schedule
✓ Procrastinate
✓ Underestimate the time it will take to conduct the EMS audit
✓ Allow an auitee to "opt out" of the audit schedule because the manager insists that his or her area is too busy to be audited.
✓ Panic when the audit doesn't go exactly to plan

FIGURE 4-3.   Planning and scheduling an EMS audit.

Document Name: XYZ Company EMS Audit Plan
Document Control No. EMS 4.5.4 C
Document Owner: Dan Johnston, EMS Coordinator
Document Approver: Mary Smith, Environmental Manager
Date: 06/03/99

Planned Audit Dates: September 21, 22, 23, 1999

Auditors Names and Qualifications:
The EMS audit team will consist of Jean Williams and John Schmidt, both of whom are independent of XYZ organization's activities, products and services and are, therefore, objective and unbiased. Specific qualifications of the audit team are listed below.

Jean Williams, ABC Consulting
1. B.S. in Civil Engineering
2. 10 years experience in environmental field
3. Knowledge of ISO 14001 Standard
4. Experienced in auditing environmental management programs

John Schmidt, ABC Consulting
1. Masters in Environmental Engineering
2. Over 15 years experience in environmental field
2. Successfully passed ISO 14001 and ISO 9000 Lead Auditor Training
3. Knowledge of ISO 14001 Standard

Scope: All elements of ISO 14001 with respect to the XYZ environmental management system. This will be a complete system audit.

Audit Process:
Opening Meeting: An opening meeting will be held with the audit team and appropriate XYZ staff. This meeting will include introduction of team members, agenda, audit logistics, and tentative schedule for closing meeting.

Conducting the Audit: The auditors will review documentation and conduct interviews with member of the XYZ as identified in audit agenda. Auditors will examine objective evidence of the selected elements of the EMS. Auditors will obtain acknowledgment of nonconformances.

Closing Meeting: A closing meeting will be held with the audit team and appropriate XYZ staff and management to discuss the results of the audit. The lead auditor must ensure that each nonconformance is understood.

FIGURE 4-4.   Example audit plan.

Document Name: XYZ Company EMS Audit Plan
Document Control No. EMS 4.5.4 C

Audit Report: The final audit report will be prepared by the lead auditor. The report is used to document the audit and summarize the findings and will include:

- names of audit team members
- dates of audit
- audit scope
- ISO 14001 elements audited
- nonconformances
- observations

Closure of Nonconformances: Any nonconformance that may arise during the EMS audit will be tracked to closure following the requirements identified in the document EMS 4.5.2.

Areas/Elements to Be Audited

Audit Topic	XYZ Personnel	Comments
Overview of XYZ EMS	EMS Management Rep., Environmental Staff	Present summary of EMS improvements.
Environmental Policy (4.2)	XYZ Top Management, EMS Management Rep.,	Review how XYZ supports the policy.
- Aspects Identification (4.3.1) - Legal & Other Reqts. (4.3.2) - Objectives & Targets (4.3.3) - Environmental Mgt. Program (4.3.4)	EMS Management Rep., Legal Counsel, Environmental Staff	Show updated significant aspects, new objectives and targets, and enhanced environmental management program.
Structure & Responsibility (4.4.1)	EMS Management Rep.	Review communication methods and interview employees
Monitoring and Measurement (4.5.1) Energy consumption	Environmental Staff -- Program engineer for energy	Review energy conservation projects and corresponding data.
Monitoring and Measurement (4.5.1) Wastewater discharges	Environmental Staff -- Program engineer for wastewater	Review training records, calibration records, and operating procedure.
Monitoring and Measurement (4.5.1) Hazardous waste discharges	Environmental Staff -- Program engineer for chemical/waste management	Review training records, operating procedures, and emergency response procedures.
Monitoring and Measurement (4.5.1) Chemical management/spills	Environmental Staff -- Program engineer for chemical/waste management	Review emergency procedures, corrective/preventive actions, and training records.

FIGURE 4-4.   *(Continued)*

Document Name: XYZ Company EMS Audit Plan
Document Control No. EMS 4.5.4 C

Areas/Elements to be Audited *Cont.*

Audit Topic	XYZ Personnel	Comments
Monitoring and Measurement (4.5.1) Air emissions	Environmental Staff -- Program engineer for air emissions	Review permits, calibration records, operating procedures, and training records.
Emergency Planning and Response (4.4.7)	Environmental Staff -- Program engineer for chemical/waste management, Security	Review emergency plan, training records, and incident records.
Monitoring and Measurement (4.5.1) HAZMAT transport	Shipping Manager	Review shipping documents and training records.
- EMS Audit (4.5.4) - Nonconformance and Corrective/ Preventive Action (4.5.2)	Environmental Staff -- EMS Audit Coordinator	Review audit plan, methodology, auditor qualification, auditor training records, and audit reports.
- Training, Awareness, and Competence (4.4.2) - Communication (4.4.3)	EMS Management Rep., Training Coordinator; Communications Coordinator	Review training needs, training records, and mechanisms for internal and external communications.
Operational Control Procedures/Records (4.4.6/4.5.3) Manufacturing Area	Manager and Personnel from Manufacturing	Interview employees about how their job can impact the environment.
- EMS Documentation (4.4.4) - Document Control (4.4.5) - Records (4.5.3)	EMS Management Rep., Environmental Staff	Review core elements of EMS and related documents; review document control procedure and test documents.
Management Review (4.6)	Top Management, EMS Management Rep.	Review how top management determines suitability, adequacy, and effectiveness of the EMS.
Operational Control Procedures/Records (4.4.6/4.5.3) Chemical/Waste Center	Manager and Personnel from Chemical/Waste Center	Review training records and operating procedures; interview employees on emergency response actions.
Operational Control Procedures/Records (4.4.6/4.5.3) Utility Plants	Manager and Personnel from the Utility Plants A, B, and C	Review training records, calibration records, and operating procedures.
Operational Control Procedures/Records (4.4.6/4.5.3) Contractors	Contractors on Premises	Interview employees about environmental policy and how their job can impact the environment.

FIGURE 4-4. (*Continued*)

XYZ's EMS Internal Audit Plan by Area/Element -- Revised 01/12/99

Audit Area	Applicable Elements	Previous Audit Date	Next Audit Date	Focus Elements	Auditor(s)
Env. Programs Staff	ALL	03 /98	02/99	ALL	Williams/Hunt
X Y Z Product Dev④	4.2, 4.3 (all). 4. 4.4.2, 4.4.3, 4.4.6, 4.4.7, 4.5.3	04/98	04/99	4.4.2②	Kirby/Sidney
X Y Z Product Testing④	4.2, 4.4.2, 4.4.3, 4.4.6, 4.4.7, 4.5.3	04/98	03/99	4.4.2②	Sidney/Hunt
X Y Z Manufacturing④	4.2, 4.3 (all) 4.4.2, 4.4.3, 4.4.6, 4.4.7, 4.5.3	04/98	02/99	4.4.2②	Williams
Admin Services④	4.2, 4.4.2, 4.4.3, 4.4.6, 4.4.7, 4.5.3	06/98	05/99	4.4.2②	Kirby
Legal Department	4.3.2	09/99	08/99	4.3.2	Hunt
Facilities Contractors	4.2, 4.4.2, 4.4.3, 4.4.6, 4.4.7	06/98	06/99	4.4.6③	Sidney/Kirby
Chemical/Waste Management Dept	4.2, 4.4.2, 4.4.3, 4.4.6, 4.4.7, 4.5.1, 4.5.3	08/98	03/99	4.4.6③	Williams/Hunt
Chemical/Waste Management Dept	4.4.6, 4.5.3	12/98	07/99	4.4.6, ③ 4.5.3 ③	Hunt
Security/Emergency Response Team	4.2, 4.4.2, 4.4.3, 4.4.6, 4.4.7, 4.5.3	08/98	09/99	4.4.7③	Kirby
Wastewater Treatment	4.2, 4.4.2, 4.4.3, 4.4.4, 4.4.5, 4.4.6, 4.4.7, 4.5.1, 4.5.3	11/98	06/99	4.4.5①, 4.4.6③, 4.5.3③	Williams/Kirby
Nonfacilities Contractors	4.2, 4.4.2, 4.4.6, 4.5.3	09/98	09/99	4.4.6, ③ 4.5.3 ③	Kirby
X Y Z Top Mgmt	4.2, 4.5.4, 4.6	04/98	05/99	4.6	Williams

NOTE: it is the responsibility of the auditor(s) to coordinate exact audit dates, scope, agenda
① focus item due to internal EMS audit results
② focus item due to other assessment results
③ focus item due to environmental importance of the activity
④ a small sample will be used for these organizations, which do not have operations of environmental importance

Note: there may be more than one applicable reason for focus on an element in an area (audit NC of an important environmental activity), but only the lowest issue number (① - ③) is listed.

FIGURE 4-5. Example audit schedule.

68

	Area to Be Audited				
1999 EMS Audit Schedule for XYZ Company	**M A N U F A C T U R E**	**F A C I L I T I E S**	**E N V P R O G R A M**	**C O N T R A C T O R S**	**T O P M A N A G M T**
ISO 14001 Requirement					
4.2 Environmental Policy	02/08 - 02/10	03/16	05/12	06/10; 12/02	04/03
4.3.1 Environmental Aspects	02/08 - 02/10	03/16	05/12		
4.3.2 Legal and Other Requirements		03/16	05/12		
4.3.3 Objectives and Targets	02/08 - 02/10	03/16	05/12		04/03
4.3.4 Environmental Management Program			05/12 - 05/13		
4.4.1 Structure and Responsibility			05/13		04/03
4.4.2 Training, Awareness and Competence	02/08 - 02/10	03/16	05/13	06/10 12/02	
4.4.3 Communication	02/08 - 02/10	03/16	05/13		
4.4.4 Environmental Management System Documentation			05/13		
4.4.5 Document Control		03/16	05/13		
4.4.6 Operational Control	02/08 - 02/10	03/16; 09/21	05/14	06/10 12/02	
4.4.7 Emergency Preparedness and Response	02/08 - 02/10	03/16; 09/21	05/14	06/10; 12/02	
4.5.1 Monitoring and Measurement		03/16; 09/21	05/14		
4.5.2 Nonconformance and Corrective/Preventive Action			05/14		
4.5.3 Records	02/10	03/16	05/14	12/02	
4.5.4 Environmental Management System Audit			05/14		
4.6 Management Review			04/03		04/03

FIGURE 4-6.   Example audit schedule matrix.

SAMPLING

Like all audits, the EMS audit is based on a sampling process because an auditor will not have time to examine all activities, operations, and documents related to the EMS. Typically, the auditor considers the size and potential environmental impact of the activity and/or function or department being audited and then randomly selects a sufficient number of samples to adequately represent the EMS. In some cases, the auditor may use a statistical sampling technique when determining the appropriate sample to be selected during a particular audit.

It is also important to remind the auditee(s) that zero nonconformance in a sample set does not mean there is no problem. Because of the sampling process, the audit may not uncover an existing nonconformance or noncompliance.

CONDUCTING AN INTERNAL EMS AUDIT

SAY WHAT YOU DO. DO WHAT YOU SAY. SHOW ME.

Once the environmental management system (EMS) audit program and procedures have been established, a complete internal EMS audit should be conducted. As mentioned in Chapter 4, depending on resources and makeup of the organization, the EMS audit can be conducted in "pieces," over a planned period of time, or it can be conducted in a single auditing session. If the audit is conducted in a single session, the auditors may want to review some EMS data prior to beginning the audit. Examples of this type of data is presented in Figure 5-1.

This chapter provides information about how to validate conformance to ISO 14001 and the organization's planned arrangement during the internal EMS audit. Each element of the ISO 14001 standard is reviewed separately and includes an audit methodology, example questions an auditor might ask when verifying conformance, and sources of objective evidence to validate conformance.

It is important to note that although checklists are a valuable tool, they should only be used as a starting point. The EMS auditor should be experienced enough to follow the flow of the

---

**Data that Might be Reviewed Before
Conducting a Complete EMS Audit**

✓ Mission statement of organization's activities, services, and
   products
✓ Physical layout of facility
✓ List of significant aspects
✓ Legal requirements/regulations that pertain to the
   organization's activities, products, and services
✓ Objectives and targets
✓ Defined roles, responsibility and authorities
✓ Core elements of the EMS
✓ Documents/procedures that support and interact with the
   core elements of the EMS
✓ Procedures for operational control
✓ Legal/regulatory compliance evaluation results
✓ Environmental records
✓ Last EMS audit results
✓ Meeting minutes of last management review

FIGURE 5-1. Examples of data that might be reviewed
before conducting a complete EMS audit.

question and answer process and not feel restricted by a check-list. See Chapter 8 for additional information on this concept.

ENVIRONMENTAL POLICY (SECTION 4.2)

AUDIT METHODOLOGY

Determine if the organization has met the requirements of the ISO 14001 standard with respect to this element by reviewing evidence that top management has defined and documented the organization's environmental policy and that the policy

- Is appropriate to the nature, scale, and environmental impacts of its activities, products, and services
- Includes a commitment to continual improvement and prevention of pollution
- Includes a commitment to comply with relevant environmental legislation and regulations and with other requirements to which the organization subscribes
- Provides the framework for setting and reviewing environmental objectives and targets
- Is documented, implemented, maintained, and communicated to employees
- Is available to the public

EXAMPLE AUDITOR QUESTIONS

Example questions asked at the relevant levels of the organization with respect to the environmental policy are presented in Table 5-1.

SOURCES OF OBJECTIVE EVIDENCE TO VALIDATE CONFORMANCE

An up-to-date and documented environmental policy, approved and defined by top management, is a "must" to show

Table 5-1. Example Auditor Questions Pertaining to the Organization's Environmental Policy

---

Questions that Could Be Asked of Top Management and/or the EMS Management Representative

- When was your environmental policy written and/or updated last?
- What is the nature of your organization's business?
- How do the elements in your environmental policy relate specifically to your business?
- How does your policy set the framework for setting objectives and targets?
- How does your policy reflect that you are committed to continual improvement?
- How has the policy been communicated to employees? To contractors?
- How is the environmental policy made available to the public?
- What do you expect employees to know about the environmental policy?
- Who are the interested parties considered when developing your environmental policy?
- How is the need for changes to the policy determined?
- How are the changes to policy communicated?

Questions that Could Be Asked of the Environmental Staff

- Are you aware of your company's environmental policy?
- How was the environmental policy communicated to you?
- How do you support the elements of your company's policy within the scope of your job?
- Can you show me evidence that demonstrates implementation of the policy at your location?
- How do you communicate the need for changes to the environmental policy?

TABLE 5-1. Example Auditor Questions Pertaining to the Organization's Environmental Policy (*Continued*)

Questions that Could Be Asked of Employees and Managers Selected at Random

- Are you aware of your company's environmental policy?
- How was the environmental policy communicated to you?
- What is your understanding of the policy?
- How is the policy implemented at your site?
- How do your responsibilities relate to the policy?
- What do the employees in your department understand of the policy (managers only)?

conformance to this element of ISO 14001 because this document is the foundation of the EMS. Examples of objective evidence that the policy has been communicated to employees, contractors, and the public are presented in Figure 5-2. Even with hard evidence in hand, the auditor will still interview employees and contractors to ascertain if the communication programs were successful.

PLANNING (SECTION 4.3)

Auditing of planning elements (Sections 4.3.1 through 4.3.4) is defined in the following paragraphs.

ENVIRONMENTAL ASPECTS (SECTION 4.3.1)

Audit Methodology.   Determine if

- The organization has established and maintained a procedure for the identification of significant aspects
- The procedure is adequate and effective

Verify that the procedure has been implemented.

Objective Evidence That Could Demonstrate that the Environmental Policy was Communicated/Made Available

Communicated To Employees / Contractors

✓ Policy is reviewed at new employee / contractor orientation
✓ Policy is reviewed during employee or contractor meetings
✓ Policy is displayed on posters, bulletin boards, and / or tent cards in the cafeteria
✓ Information about policy is provided in employee newsletters
✓ Policy is included in work contracts
✓ Policy is posted near paper and / or aluminum can recycling areas
✓ Information about the policy is provided during Earth Day activities
✓ Policy is posted in heavily trafficked areas, such as in the cafeteria or in the copier / FAX area

Made Available To the Public

✓ Policy is published in annual reports
✓ Policy is published in press announcement
✓ Policy and environmental achievements are presented at neighborhood meetings
✓ Policy is available on company WEB page

FIGURE 5-2. Objective evidence that could demonstrate that the environmental policy was communicated or made available.

Example Auditor Questions.   Example auditor questions asked at the relevant levels of the organization with respect to the environmental aspects are presented in Table 5-2.

Sources of Objective Evidence to Validate Conformance. There is no "right way" to evaluate aspects and significant aspects. Some organizations chose to use a quantitative technique, whereas others use a qualitative means of making the evaluation. The key point here is to have a procedure—preferably documented—and to follow it. An example of a procedure for an evaluation process that utilizes a qualitative technique and a sample evaluation performed according to the procedure is presented in Figure 5.3(*a*) and (*b*). An example procedure for an evaluation process that utilizes a quantitative technique and a sample evaluation performed according to the procedure is presented in Figure 5.4(*a*) and (*b*). As can be seen from the procedures, the criteria and/or rationale used in the evaluation process is included. This criteria should clearly indicate what factors were considered to determine significance. After significant aspects are identified, employees whose jobs can affect these aspects should be informed of their role with respect to these significant aspects and the EMS process.

Sources of objective evidence to demonstrate conformance to this element might include

- The procedure used for identifying environmental inputs, outputs, and significant aspects
- A list of all aspects that apply to your location
- A list of significant aspects that apply to your location and rationale for why these were selected as significant (or why other were not selected as significant)
- Meeting minutes of the aspect evaluation process
- Interviews with employees whose jobs could impact significant aspects to ensure they understand their role in the EMS process
- Interviews with management and employees involved with aspect and/or significant aspect identification.

TABLE 5-2. Example Auditor Questions Pertaining to the
Organization's Environmental Aspects and Significant Aspects

Questions that Could Be Asked of the EMS Management Representative and/or Environmental Staff

- What are the inputs and outputs at your location?
- What are the environmental aspects or impacts of your location?
- How did you identify the environmental aspects?
- What procedure did you use to determine significant environmental aspects?
- What criteria was used to determine which aspects are significant?
- For those aspects that were not considered to be significant, explain why not. (Note: This is not required by the standard but will relate to the aspects evaluation process.)
- Provide objective evidence (meeting minutes, criteria charts, assessment documents) that the aspects evaluation took place according to the procedure.
- Provide the list of significant aspects.
- How do these significant aspects relate to objectives and targets? (Note: Objectives and targets do not have to be established for all significant aspects; some of the significant aspects may be under operational control.)
- How did you consider impact on community and views of interested parties?
- How and when do you reevaluate aspects and significant aspects?
- How are significant aspects communicated to relevant organizations? (Note: Any answer you give will be tested.)

Questions that Could Be Asked of Employees and Managers Selected at Random

- What are the environmental aspects or impacts of your job?
- What procedures do you have in place to control environmental impacts of your job?

Procedure Name: Identifying Significant Environmental Aspects
Document Control Number: EMS Procedure 4.3.1
Document Owner: Mary Smith, Environmental Manager, XYZ Company

The Environmental Manager of XYZ Company initiates the process of identifying environmental aspects of activities and services. The process includes soliciting input from professionals from manufacturing, facilities, procurement, and distribution. The following information is considered:

a) inputs and outputs from routine operations;
b) inputs and outputs from major maintenance and/or turnaround activities; and
c) potential for accidents and emergency situations and their effect on the environment.
d) inputs and outputs from services that could significantly affect the environment

The determination of significant environmental impacts will be based on the best judgment of the professionals involved in this process, with advice and counsel from other experts, as needed. The determination will consider, at a minimum:

a) legal/regulatory requirements pertaining to activities, products, and services;
b) risk to employees and/or neighborhood populations;
c) environmental impact;
d) public perception, including customer views.

Environmental aspects which have or can have significant environmental impacts are classified as significant environmental aspects. Significant environmental aspects are documented, reviewed, and updated, as necessary. At a minimum, significant environmental aspects are reviewed annually.

The Environmental Manager communicates the list of XYZ's significant environmental aspects to the XYZ management team and relevant personnel. This communication takes place after the initial identification of significant environmental aspects and whenever there is a change in these. If there are no changes within a calendar year, then the Environmental Manager confirms this to XYZ's management team at the end of the calendar year.

FIGURE 5-3a.   Example procedure for determining significant aspects.

Significant Environmental Aspects Criteria	L	R	EI	P	Significant?
Activities and Services					
1. Energy consumption	N	N	Y	Y	Yes
2. Water consumption	N	Y	Y	N	No
3. Wastewater discharges	Y	Y	N	N	No
4. Hazardous waste generation	Y	Y	Y	Y	Yes
5. Nonhazardous waste generation	N	Y	Y	N	Yes
6. Chemical spills	Y	Y	Y	Y	Yes
7. Air emissions	Y	Y	Y	Y	Yes
8. Chemical management	Y	Y	Y	Y	Yes
9. Noise	Y	N	N	N	No
10. Hazardous materials transport	Y	Y	Y	Y	Yes
11. Nonhazardous materials transport	N	N	Y	N	No
Products					
12. Size / Weight	N	N	Y	N	No
13. Recyclable / Reusable	N	N	Y	Y	Yes
14. Energy consumption	N	Y	Y	Y	Yes
15. Recycled plastics content	N	N	Y	Y	Yes

Rationale for Determining Significance

Activities and Services

1. Energy consumption has significant environmental focus because of the environmental impact created during its generation, including acid rain, air pollution, climate change, and ozone depletion.
2. Water consumption in the area that XYZ operates has minimal environmental impact because there is an abundant water supply and the major water use is domestic use.
3. Wastewater discharges are comprised of nonhazardous, domestic discharges. Local wastewater treatment is designed to accommodate the types and volume of wastewater flow. No chemical waste streams are discharged into the wastewater system.
4. Hazardous waste generation is an environmental focus in terms of waste minimization and proper handling, treatment, and disposal of this waste.
5. Nonhazardous waste is significant because of the large volumes generated from the packaging and shipment of XYZ products.
6. Chemical spills -- that is, unintended releases to the environment -- could potentially harm air, land, and / or water, and must be managed carefully using proper procedures should they occur.
7. Chemicals must be managed carefully in terms of use, handling, distribution (transport), and discharge.
8. Air emissions from manufacturing or facilities operations can impact air quality and surrounding neighbors.
9. Environmental noise from XYZ is not a factor because the type of manufacturing and facilities operations are relatively noise-free, and there is a 7-acre buffer zone around the facility.
10. There could be a significant environmental impact from hazardous materials transport such an accident or emergency incident occur.
11. There is minimal risk of environmental impact from transport of XYZ's nonhazardous waste.

Products

12. Size/weight considerations with respect to XYZ's products are not a major focus since the products are small.
13. Recyclability/reusability is focused upon within XYZ's product takeback department.
14. Product energy consumption is a key customer concern, which makes this product aspect significant.
15. Recycled plastics content in XYZ's products has become a focus item because of resource use and customer perception.

FIGURE 5-3b.   Matrix for determining significance based on qualitative criteria.

Procedure Name: Identifying Significant Environmental Aspects
Document Control Number: EMS 4.3.1
Document Owner: Mary Smith, Environmental Manager

The Environmental Manager of XYZ Company initiates the process of identifying environmental aspects of activities, products, and services. The process includes soliciting input from professionals from manufacturing, facilities, procurement, and distribution. The following information is considered:

a) inputs and outputs from routine operations;
b) inputs and outputs from major maintenance and/or turnaround activities; and
c) potential for accidents and emergency situations and their effect on the environment.
d) inputs and outputs from services that could significantly affect the environment

The determination of significant environmental impacts will be based on the numeric process defined in Attachment A of this procedure, with each expert evaluating each aspect. The determination will consider, at a minimum:

a) legal/regulatory requirements pertaining to activities, products, and services;
b) risk to employees and/or neighborhood populations and/or customers;
c) environmental impact frequency;
d) environmental impact;
e) public perception, including customer views.

Environmental aspects which have or can have significant environmental impacts are classified as significant environmental aspects. Significant environmental aspects are documented, reviewed, and updated, as necessary. At a minimum, significant environmental aspects are reviewed annually.

The Environmental Manager communicates the list of XYZ's significant environmental aspects to the XYZ management team and relevant personnel. This communication takes place after the initial identification of significant environmental aspects and whenever there is a change in these. If there are no changes within a calendar year, then the Environmental Manager confirms this to XYZ's management team at the end of the calendar year.

FIGURE 5-4a.   Example procedure for determining significant aspects.

Procedure Name: Identifying Significant Environmental Aspects
Document Control Number: EMS Procedure 4.3.1

ATTACHMENT A   Environmental Aspect Identification Process

Criteria Used/Possible Ratings

1. Legal/Regulatory Requirements (L) -- Is there a legal/regulatory requirement or a permit required?
   0 = There is no legal/regulatory requirement
   3 = There is a legal/regulatory requirement
   5 = There is a permit required

2. Risk (R) - Rate the potential risk to employees and/or neighbor populations.
   1 = Low Risk
   3 = Intermediate Risk
   5 = High Risk

3. Environmental Impact Frequency (F) - Rate the frequency of occurrence.
   1= Low Frequency
   3 = Intermediate Frequency
   5 = High Frequency

4. Environmental Impact (EI) - Classify the impact according the importance.
   1 = Low Importance
   3 = Intermediate Importance
   5 = High Importance

5. Public Perception (P) - Determine the importance of the environmental impact in terms of public perception.
   1 = Low Perception
   3 = Intermediate Perception
   5 = High Perception

Criteria for determining significance of the aspect:

1. Aspect where the sum of values is > or = 15.
2. Aspect where (F+ EI) > 6
3. Aspect where   P = 5

FIGURE 5-4a. (*Continued*)

Significant Environmental Aspects Criteria	L	R	F	EI	P	Significant?
Activities and Services						
1. Energy consumption	0	1	5	3	3	Yes
2. Water consumption	0	1	5	1	1	No
3. Wastewater discharges	3	1	5	3	3	Yes
4. Hazardous waste generation	5	3	3	5	3	Yes
5. Nonhazardous waste generation	3	1	5	1	3	No
6. Chemical spills	3	5	1	5	5	Yes
7. Air emissions	5	1	5	3	3	Yes
8. Chemical management	3	5	5	3	3	Yes
9. Noise	0	1	3	1	1	No
10. Hazardous materials transport	5	3	3	5	3	Yes
11. Nonhazardous materials transport	3	1	3	1	1	No
Products						
12. Size / Weight	0	0	3	3	3	No
13. Recyclable / Reusable	0	0	3	5	3	Yes
14. Energy consumption	0	3	5	3	5	Yes
15. Recycled plastics content	0	0	5	3	5	Yes

FIGURE 5-4b. Matrix for determining significance based on numeric criteria.

LEGAL AND OTHER REQUIREMENTS (SECTION 4.3.2)

Audit Methodology.   Determine if the organization has a procedure for identifying and providing access to legal requirements and other voluntary requirements to which the organization subscribes. Validate that the procedure has been implemented and that relevant personnel have access to legal and other requirements to which the organization subscribes. Validate that the organization has the necessary permits for its activities, products, and services and that the permits are being met.

Example Auditor Questions.   Example questions asked at the relevant levels of the organization with respect to legal and other requirements are presented in Table 5-3.

Sources of Objective Evidence to Validate Conformance. Although not required by the standard, it is best to have a high-level list of applicable laws and regulations. This demonstrates to the auditors that the organization knows exactly which legal and regulatory requirements apply to them. In addition, this can serve as a valuable tool for training new employees and for providing employees with a central place to access this information. Finally, the required procedure for identifying legal and regulatory requirements and for providing access to those whose jobs could affect these requirements should be readily available. Example methods for identifying legal and regulatory requirements and for communicating these requirements to those that need to know are presented in Figure 5-5.

OBJECTIVES AND TARGETS (SECTION 4.3.3)

Audit Methodology.   Determine if the organization considered the elements required by ISO 14001 standard when setting objectives and targets. Validate that the organization has established objectives and targets and that the following elements were considered by the location when setting these:

- Legal and other requirements
- Significant environmental aspects

TABLE **5-3.** Example Auditor Questions Pertaining to the
Organization's Legal and Other Requirements

---

*Questions Asked of the EMS Management Representative, Environmental Staff
and/or Legal Staff*

- What is your procedure to identify legal and other requirements
  to which the organization subscribes?

- What laws and other requirements apply to your environmental
  aspects?

- What are the sources of legal information?

- How do you keep this information current?

- How do you decide whether a pending or new legal requirement requires attention?

- How do you communicate legal and other requirements to
  relevant organizations or employees?

- Do you subscribe to any voluntary commitments?

- What are the responsibilities of your location legal department?

- How do you ensure the information is current?

*Questions Asked of Employees and Managers Selected at Random (Whose Job
Activities Can Impact Legal Requirements—i.e., Waste Handlers and
Wastewater Treatment Operators)*

- How do you know the legal requirements of your job?

- Where can you find this information?

- How do you communicate them to relevant personnel?

- What environmental permits do you have? (Note: Compliance
  to permit requirements will be tested.)

- How do you know what permits are needed?

- Who prepares the permit application?

- Who is authorized to sign the permit application?

- What laws and other requirements apply to your environmental aspects?

- If there are changes to the laws, how is this communicated
  to you?

**Examples of How to Obtain Legal/Regulatory
Requirements and Provide Access to Relevant Employees**

Obtaining Legal/Regulatory Requirements

✓ Permits and/or Consent Orders
✓ On-line publications that pertain to the organization's
   activities, products, and services
✓ Hard copy publications that pertain to the organization's
   activities, products, and services
✓ Government/agency newsletters
✓ Technical societies
✓ Technical publications
✓ Consultants

Providing Access to Legal/Regulatory Requirements

✓ Posting of permits and/or consent orders
✓ Training
✓ Operating procedures
✓ Work instructions
✓ Emergency response procedures

FIGURE 5-5.   Examples of how to obtain and provide
access to legal/regulatory requirements.

- Technological options
- Financial requirements
- Business requirements
- Views of interested parties

Example Auditor Questions.   Example questions asked at the relevant levels of the organization with respect to objectives and targets are presented in Table 5-4.

Sources of Objective Evidence to Validate Conformance. The first source of objective evidence to validate conformance to this element of the standard is to have a list of objectives and targets. An objective does not have to have a numeric target, unless it is practicable to set one. In addition to the list of objectives and targets, the organization should have a rationale for how it used the required criteria when setting them. Examples of objectives and targets that might be considered by an organization and the rationale for inclusion or exclusion are presented in the case studies in Figure 5-6.

Environmental Management Program (Section 4.3.4)

Audit Methodology.   Determine if the organization has in place and is following an environmental management program that meets the requirements of ISO 14001 and that supports the objectives and targets set for activities, products, and services. Validate that the organization's environmental management program is current and being followed and that it includes the following with respect to achieving objectives and targets:

- Time frame for achieving objectives and targets
- Person responsible for activities related to achieving objectives and targets
- Means for achieving objectives and targets

Example Auditor Questions.   Example questions asked at the relevant levels of the organization with respect to the environmental management program are presented in Table 5-5.

TABLE 5-4. Example Auditor Questions Pertaining to the
Organization's Objectives and Targets

---

*Questions that Could Be Asked of the EMS Management Representative and/or
Environmental Staff*

- What are your location's objectives and targets?
- Have you set objectives and targets for all significant aspects?
  (Note: This is not a requirement of the standard, but it may
  be asked.)
- What criteria was used to establish the objectives and targets?
- At what level (relevant function) are the objectives and tar-
  gets set and what is the rationale for this?
- How were the following considered when setting objectives
  and targets:

    Legal and other requirements

    Significant environmental aspects

    Technological options

    Financial, operational, and business requirements

    Views of interested parties

- How do you ensure that objectives and targets are consistent
  with the commitment of continual improvement?
- How are objectives and targets communicated?
- How are objective and targets monitored?
- How often do you review your objectives and targets to assess
  if they are current and/or if new ones need to be set?

Questions Asked of Employees and Managers Selected at Random

- What are your location's objectives and targets as they relate
  to your job?
- Other than the site objective and targets, does your department have any specific objectives and targets?
- What procedure or programs do you have in place to meet
  objectives?
- How do you monitor and measure and report progress toward
  objectives and targets?

Case Study 1 for Setting Objectives and Targets

An airline company identified chemical use, specifically use of deicing materials, as a significant aspect. When setting objectives and targets the following information was considered for the use of deicing materials:

- Legal requirements -- There are some legal requirements applicable in specific locations in which the company operates.
- Significant environmental aspect -- As an element of chemical use, it was identified as a significant aspect.
- Technological options -- The materials are difficult to capture and treat. There may be materials that are less toxic than those that the company is currently using.
- Financial requirements -- Very expensive to treat. An evaluation of alternative materials could be performed by a consultant for approximately $10,000. Reformulation studies an be conducted by chemical laboratory, at a cost of approximately $50,000.
- Business requirements -- Deicing materials are essential to the business.
- Views of interested parties -- The perception is that the deicing materials may cause harm to plants and wildlife. However, all parties understand that these materials are needed for safe functioning of planes during winter months.

Based on the above information, the organization decided to establish the following objective pertaining to use of deicing materials:

Objective: Initiate program to identify an environmentally safe deicing material.

Target: No target was set for this objective because it is impracticable to do so at this time.

FIGURE 5-6a.   Case study 1 for setting objectives and targets.

Case Study 2 for Setting Objectives and Targets

A company which manufactures radios uses a large quantity of plastic. The company identified the use of plastics as a significant aspect. When setting objective and targets the following information was considered with respect to use of plastics and consumption of raw material:

- Legal requirements – There are no legal requirements governing the use of plastics and the consumption of raw material.
- Significant environmental aspect -- use of plastics and consumption of raw material were identified as a significant aspects.
- Technological options: Options include making radios smaller, using alternate materials, or making radios with some percentage of recycled plastic
- Financial requirements - Alternate materials are very expensive
- Business requirements - Use of plastics is essential to business competitiveness
- Views of interested parties - Plastic is non-biodegradable; some interested parties want to return radios to the manufacturer instead of discarding them through public landfill system

When setting objectives and targets, the organization felt it had opportunity to use a technological option to minimize the impact of the aspect and established the following objective and targets:

Objective: Increase the percentage of recycled plastic content in new radios

Targets: (1) By year-end 1999, all radios will have a recycled plastics content of at least 20%; (2) By year-end 2000, all radios will have a recycled plastics content of at least 50%; (3) By year-end 2001, all radios will have a recycled plastics content of at least 70%.

In addition, the organization felt that customer views warranted consideration, and the following objective and target was established:

Objective: By year-end 1999, a program will be established to accept returned (unwanted) radios from customers

Target: By year-end 2000, the organization will recycle 20% of all radios sold

Note: Specifics for this objective and target are identified in the environmental program)

FIGURE 5-6b.   Case study 2 for setting objectives and targets.

TABLE 5-5. Example Auditor Questions Pertaining to the
Organization's Environmental Management Program

Questions that Could Be Asked of the EMS Management Representative and/or Environmental Staff

- What procedure or programs do you have in place to meet objectives and targets?

- Show me your environmental management program? (Note: This program must include objective; target, if applicable; person responsible; implementation means/methods; and time frame.)

- How do you communicate the environmental management program to relevant personnel?

- What do you do if you don't meet the established objectives and targets?

- How would you amend the program if there are new developments that can affect the objectuves and targets?

- Verify that resources are identified (human and/or financial).

Questions Asked of Employees and Managers Selected at Random

- How have your organization's objectives and targets been communicated to you?

- What procedures or programs do you have in place to meet your site's objectives and targets?

- How do you measure your department's progress toward meeting objectives and targets at your level, and do you report this progress to anyone?

- What do you do if you fall behind on meeting your objectives and targets?

Sources of Objective Evidence to Validate Conformance.
Validating conformance to this element will include having an environmental management program that specifies the objective, target, person responsible for achieving (or coordinating the achievement of) the objective and target, the means and method of achievement, and the time frame. Example environmental management programs for the objectives and targets set in case studies depicted in Figure 5-6 are presented in Figure 5-7.

Environmental Management Program for Case Study 1

Objective: Initiate program to identify an environmentally safe deicing material.

Target: None was set for this objective because it is impracticable to do so at this time.

Responsible Person: Jon Bulgrat, Department 452, Environmental and Safety Engineering

Time frame: By year end an environmentally safe deicing material will be identified

Means:

(1) Retain consulting firm with toxicological expertise to perform literature search for environmentally safe materials that have deicing capabilities equal to or better than currently used material. (1Q)

(2) If unsuccessful in finding "off the shelf" alternative, hire chemical laboratory to review reformulation of currently used deicing materials. (2Q-3Q)

(3) Allocate budget as follows: $10,000 for consultant fees; $50,000 for laboratory fees. (1Q)

(4) Hold monthly meetings with the environmental staff, engineering department, and safety department to discuss progress. (Ongoing throughout year)

FIGURE 5-7.   Example environmental management program.

Environmental Management Program for Case Study 2

Objective: Increase the use of recycled plastic content in new radios.

Targets: (1) By year-end 1999, all radios will have a recycled plastics content of at least 20%; (2) By year-end 2000, all radios will have a recycled plastics content of at least 50%; (3) By year-end 2001, all radios will have a recycled plastics content of at least 70%.

Responsible person: Carol Jenko, Manager, Product Development Staff.

Time frame: Refer to target.

Means for Target (1): Identify companies that recycle plastic; evaluate recycled plastic composition to ensure product suitability (1Q'99). Select company to supply recycled plastic; finalize purchasing agreements; work with development and initiate prototype (2Q'99). Prepare production to use 20% recycled material (3Q'99). Ensure all radios have 20% recycled plastic content (Year-end '99)

Means for Targets (2) and (3): Evaluate feasibility of making these targets; put plan in place for meeting targets (Year-end '99).

Objective: Minimize waste by establishing a return and recycling program for radios.

Targets: (1) By year-end 1999, the organization will establish a plan to accept returned (unwanted) radios from customers (2) By year-end 2000, the organization will recycle 50% of the plastic of all radios returned.

Responsible Person: April Anderson, Product Distribution Manager for Target (1); Stan Kelley, Procurement Manager and Vendor Liason for Target (2)

Time frame: Refer to target.

Means for Target (1): Select members of working group to develop plan to accept returned radios (2Q '99). Complete written plan (Year-end '99). Implement plan (1Q ' 2000).

Means for Target (2): Review and select vendors who can recycle the plastics from radios (3Q'99); Begin recycling program (Year-end '99). Execute target (Year-end 2000).

FIGURE 5-7.   (*Continued*)

IMPLEMENTATION AND OPERATION (SECTION 4.4)

Auditing of the implementation and operation sections of ISO 14001 (Sections 4.4.1 through 4.4.7) is discussed below.

STRUCTURE AND RESPONSIBILITY (SECTION 4.4.1)

Audit Methodology.   Determine if environmental management roles, authorities, and responsibilities have been defined by the organization. Validate that

- Roles and responsibilities have been documented and test that they have been communicated to employees at all relevant levels
- A person has been designated as the EMS management representative responsible to ensure the EMS is established, implemented, and maintained
- That the EMS management representative is responsible for coordinating or reporting on the performance of the EMS to top management for review and as a basis for continual improvement of the EMS
- Management has provided resources essential to the implementation and control of the EMS

Example Auditor Questions.   Example questions asked at the relevant levels of the organization with respect to structure and responsibility are presented in Table 5-6.

Sources of Objective Evidence to Validate Conformance. To show the auditor evidence that roles and responsibilities have been defined and documented, the organization should have an organization chart and/or other document(s) that describe the roles and responsibilities of all involved with the EMS. A sample document defining roles and responsibilities is presented in Figure 5-8.

Communication of roles and responsibilities may take several forms, such as (1) letters of delegation from top management for certain job responsibilities, (2) verbal communication—in

TABLE 5-6. Example Auditor Questions Pertaining to the Organization's Structure and Responsibility

Questions Asked of the EMS Representative

- What is your organizational structure?

- Who is top management with respect to the EMS?

- Show me the documented environmental management roles, authorities, and responsibilities for establishing, implementing, and maintaining the EMS (Note: Auditor should look for the words *has authority* and *has responsibility* and for the terms *establishes, implements,* and *maintains.* These terms should be used when describing the organization's structure and responsibilities.)

- Show me the description of your responsibility for coordinating or reporting on the performance of the EMS to top management for review and as a basis for continual improvement of the EMS.

- Who has authority to provide financial and human resources?

- Who is responsible to establish, maintain, and implement the EMS?

- Who is the appointed management representative?

- How do you communicate roles and responsibilities?

- How do you communicate changes to roles and responsibilities?

- What are the roles and responsibilities of contractors?

- How are roles and responsibilities communicated to contractors?

- How are human resources allocated and who determines that they are adequate to implement and maintain the EMS?

- How are financial resources provided for implementation and maintenance of the EMS?

- Who determines technology needs for the EMS?

TABLE 5-6. Example Auditor Questions Pertaining to the
Organization's Structure and Responsibility (*Continued*)

Questions Asked of the Environmental Staff

- What are your role(s) and responsibilities with respect to the EMS?
- How were these role(s) and responsibilities communicated to you?

Questions Asked of Employees/Managers Selected at Random

- How are responsibilities for the EMS communicated to you?
- If you needed to know who was responsible for a certain element of the EMS, how would you find out?

which case the auditor would verify this through interviews, (3) communication through job descriptions, or (4) definition of roles and responsibilities within the EMS manual, if the organization chooses to establish one. In addition, the organization must provide objective evidence that resources (including human resources and specialized skills, financial resources, and technology) are provided to implement and maintain the EMS. This type of evidence could include a skills assessment of the environmental organization, capital plans, expense plans, and other such evidence.

TRAINING, AWARENESS, AND COMPETENCE (SECTION 4.4.2)

Audit Methodology.   Determine if

- Training needs are identified and personnel whose work may create a significant impact upon the environment have received appropriate training
- There is a procedure for awareness training at each relevant function and level

Validate that

- Training needs have been identified and provided
- There is a procedure to make employees aware of the EMS

Document Name: Definition of EMS Roles and Responsibilities at XYZ Company
Document Control No. EMS Document 4.4.1
Document Owner: Mary Smith, Environmental Manager
Approver: Jim O'Brien, XYZ General Manager
Date: 02/19/99

The roles, responsibilities and authorities XYZ's environmental management system are defined as follows:

XYZ General Manager

- Has authority and responsibility to approve and issue XYZ Company's environmental policy
- Responsible for overall environmental compliance
- Has the authority to review and assess the suitability, adequacy, and effectiveness of the EMS
- Has authority to delegate specific responsibility within the EMS

EMS Management Representative

- Has authority, delegated by the General Manager to establish, implement, and maintain the EMS for the XYZ Company
- Has authority to provide the appropriate resources for implementation and maintenance of the EMS
- Responsible for reporting on the performance of the EMS to the General Manager
- Responsible for approving objectives and targets based on recommendation from the EMS core team

Manager of XYZ Environmental Department

- Responsible for identifying training requirements of the environmental staff
- Responsible for approving the XYZ EMS Manual
- Responsible to assign resources to support the EMS

FIGURE 5-8.   Sample document identifying roles and responsibilities.

Document Name: Definition of EMS Roles and Responsibilities at XYZ Company, *Cont.*
Document Control No. EMS Document 4.4.1

EMS Coordinator

- Responsible for establishing, implementing, and maintaining the EMS under the guidance of the EMS Management Representative
- Responsible for developing the ISO 14001 awareness training for XYZ employees with respect to the environmental Affairs policy and key elements of the EMS
- Responsible for organization information on the performance of the EMS and presenting it to the EMS Management Representative
- Responsible for coordinating efforts for EMS audits
- Responsible for providing guidance and input to Communication Manager for responding to relevant external communications from interested parties related to the EMS

ISO 14001 EMS Core Team

- Responsible for identifying XYZ's aspects and determining significant aspects
- Responsible for recommending objectives and targets to the EMS Management Representative

Communications Manager

- Responsible and authorized to establish, implement, and maintain sitewide communication vehicles
- Responsible for external communications/interfacing with the public (to receive and document relevant communications from external parties)

FIGURE 5-8.   (*Continued*)

Document Name: Definition of EMS Roles and Responsibilities at XYZ Company, *Cont.*
Document Control No. EMS Document 4.4.1

Department Managers

- Responsible for identifying training needs
- Responsible for establishing and maintaining operating procedures necessary for maintaining operation control of department activities
- Responsible for investigating nonconformances and for initiating preventive and corrective action applicable to the department
- Responsible for making employees aware of the affect their actions have or potentially can have on the environment
- Responsible for retaining self-audit records (according to applicable record retention requirements)
- Authorized and responsible for employee evaluations of competency

Environmental Staff

- Responsible for developing, maintaining, and implementing programs to ensure compliance with regulatory requirements, and assessing overall compliance
- Responsible for coordinating monitoring and measurement efforts for key environmental characteristics
- Responsible for external reporting as required by regulation and/or permit conditions
- Responsible for assisting the Environmental Manager to provide guidance and input to Communication Manager for responding to relevant external communications from interested parties related to XYZ's environmental management system
- Responsible for implementing, establishing, and maintaining programs to enable XYZ to meet the defined objectives and targets
- Responsible for tracking progress toward achieving objectives and targets

Security Operations/Emergency Planning Manager

- Responsible for developing and implementing program for emergency preparedness and response to environmental incidents
- Responsible for providing training to employees on emergency response
- Authority to solicit help from external organizations during an emergency

FIGURE 5-8.   *(Continued)*

Document Name: Definition of EMS Roles and Responsibilities at XYZ Company, *Cont.*
Document Control No. EMS Document 4.4.1

Emergency First Response Personnel

*   Responsible for responding to chemical emergency incidents
*   Responsible for advising Security Operations/Emergency Response Manager of corrective action and preventive action taken during incident

Site Contractors and Vendors

*   Responsible for adherence to XYZ's environmental policy and operating procedures, that are applicable to their jobs.

XYZ Personnel

*   Responsible for adherence to XYZ 's enironmental policy and procedures applicable to their jobs
*   Responsible for assisting site in meeting objectives and targets through completing actions as defined by the environmental staff
*   Responsible to be aware of the environmental impact of their job

FIGURE 5-8.   (*Continued*)

Test that employees and contractors at each relevant level and function understand the following:

- The importance of conformance with the environmental policy and procedures and with the requirements of the EMS as it pertains to them
- The significant environmental impacts, actual or potential, on their work activities and the environmental benefits of improved personal performance
- Their roles and responsibilities in achieving conformance with the environmental policy and procedures and with the requirements of the EMS, including emergency preparedness and response requirements
- The potential consequences of departure from specified operating procedures

Verify that employee competence is determined based on education, training, and/or experience.

Example Auditor Questions.   Example questions asked at the relevant levels of the organization with respect to training, awareness, and competence are presented in Table 5-7.

Sources of Objective Evidence to Validate Conformance. Sources of objective evidence to validate that training, awareness, and competence requirements have been met might take several forms. Interviews with management may be used to ascertain what is used to determine competence (i.e., number years on job, education, skills and experience history, on-the-job training, and other training). Subsequent interviews with employees may also be in order to verify that they have the needed training, education, and experience.

In addition, training matrices can provide evidence that training needs have been identified and the corresponding records show that the training has been completed. In some cases training records may be kept in a centralized database, whereas other organizations may keep these records at the department level or even at the employee level. Finally, for professional staff, training and awareness might be accomplished through technical meetings and conferences.

TABLE 5-7. Example Auditor Questions Pertaining to the
Organization's Training, Awareness and Competence

---

*Questions Asked of the EMS Management Representative and/or
Environmental Staff*

- How do you identify training needs?
- How do you make employees aware of your EMS?
- How do you keep top management aware of your EMS?
- Show me training records (including contractors and subcontractors in those areas that can have significant environmental impact).
- What are the training needs for EMS auditors?
- What type of training do your contractors need?
- How do you identify legally required training?
- How do you ensure that contractors and subcontractors on site have adequate training? (Note: Review contracts if necessary to see what training is specified.)
- How are employees determined to be competent for their jobs?

Questions Asked of Managers Selected at Random

- Are you aware of the EMS?
- How do you make employees aware of your EMS?
- What type of training do your employees need to perform and be competent in their jobs?
- How did you receive your training?
- Show me your training records.

Questions Asked of Employees Selected at Random

- Are you aware of the environmental management system?
- What type of training do you need to perform your job?
- Is there any legally required training?
- How did you receive your training?

Part of training needs identification will include training that is required legally. Examples of training that might be legally required are presented in Figure 5-9.

Communication (Section 4.4.3)

Audit Methodology.   Determine that the procedure(s) for internal communication and external communications from interested parties about the EMS is (are) established and implemented. Validate that there is a procedure(s) established for

- Internal communications at all relevant levels within the organization
- Receiving, documenting, and responding to relevant communication from external interested parties

Test that the procedure is being implemented.

Example Auditor Questions.   Example questions asked at the relevant levels of the organization with respect to communication are presented in Table 5-8.

Sources of Objective Evidence to Validate Conformance. Some examples of objective evidence to validate that the organization has met the requirements for this section pertaining to internal communications include

- Maintaining communication distribution lists
- Maintaining notes and memoranda
- Providing internal communications through the internal website, bulletin boards, and newsletters

It is important for the organization to define what is relevant with respect to external communication because it is relevant external communications that need a procedure for receiving, documenting, and responding to them. A typical way to provide evidence that this requirement is being met is to maintain an external communications log, an example of which is presented in Figure 5-10.

Examples of Legally Required Training

✓ Hazard communication training about chemicals in the
  workplace
✓ Hazardous waste management training for employees who
  handle and manage hazardous waste
✓ Wastewater treatment operator training for employees who
  work at the wastewater treatment plant
✓ Emergency response training for employees responding to
  chemical incidents
✓ Hazardous materials transportation training for employees to
  ship hazardous materials
✓ Asbestos management training for employees who remove
  asbestos-containing materials from the workplace
✓ PCB management training for employees who handle PCBs
✓ Import/export training for those who receive or ship hazardous
  materials

FIGURE 5-9. Examples of training that might be legally
required.

TABLE 5-8. Example Auditor Questions Pertaining to the Organization's Communications

---

Questions Asked of the EMS Management Representative and/or Environmental Staff

- Show me your procedure for internal and external communication.

- How do you communicate requirements of your EMS to employees and relevant organizations?

- How do you communicate your environmental policy, objectives, and targets?

- If you received a question from someone external, how would you handle it?

- How do you handle complaints from external interested parties?

- What are the criteria to determine what is relevant communication from interested parties?

- How do you decide whether to respond to an external inquiry?

- How do you communicate externally about significant aspects?

- Who receives and responds to inquiries from legal authorities?

- How do you track or log external communications or the decision not to respond to them?

- Show me examples of the above types of communications.

Questions Asked of Employees and Managers Selected at Random

- How have you received information relative to the roles and responsibilities of your job as they relate to the EMS?

- How have you received information about the environmental policy?

- How have you received information about the organization's significant aspects as they relate to your job?

- How have you received information about the organization's objectives and targets as they relate to your job?

- If you have a question relating to environmental issues, to whom and how would you communicate them?

XYZ's External Communication Log*

Communication	Received	Response	Comments
Letter from neighbor concerned about visible air emissions.	02/18/98	Replied that emissions were condensate from cooling towers.	Letter sent via certified mail on 02/25/98.
Letter from neighbor thanking us for hosting an earth day celebration for the community	04/26/98	Acknowledged neighbor's comments with return letter.	Letter sent regular mail 04/30/98.
Letter from City asking for XYZ to participate in voluntary program during ozone action days.	02/15/99	Replied concurrence to participate.	Letter sent certified mail on 02/28/99.

* Records of inquiries and responses in File C, Communication Manager's office.

FIGURE 5-10.   Example of a communication log.

EMS Documentation (Section 4.4.4)

Audit Methodology.   Determine if the core elements of the organization's EMS and documents that interact with the core elements have been identified. (Note: Core elements of the EMS are those documents integral to the EMS such as the environmental policy, procedures required by the standard, and the organization's EMS manual—should the organization choose to establish one; documents that interact with the core elements are other related documents and procedures for implementing and maintaining control of the EMS.)

Auditor Questions.   Example questions asked at the relevant levels of the organization with respect to EMS documentation are presented in Table 5-9.

Sources of Objective Evidence to Validate Conformance. For objective evidence to show that the requirements of this section have been met, the organization might want to have a list that identifies the core elements of the EMS and related documents and procedures. Examples of these two types of documents are presented in Figure 5-11.

Document Control (Section 4.4.5)

Audit Methodology.   Determine if the organization has a procedure to ensure that documents are controlled. Validate that a procedure exists to control documents. Throughout the course of the audit, test documents at point of use to validate the following attributes:

- They have proper approval.
- They are current.
- They are legible.
- They are identifiable.
- They are dated (with revision dates available).
- They are maintained in an orderly manner.
- Obsolete documents are removed promptly.

TABLE 5-9. Example Auditor Questions Pertaining to the
Organization's EMS Documentation

*Questions Asked of the EMS Management Representative and/or
Environmental Staff*

- What are the core elements of your organization's EMS?

- Who has access to these documents?

- How do you communicate the core elements of the EMS
  (Note: Specific communication of the core elements is not
  required.)

- What are the documents that support these core elements?

Questions Asked of Employees and Managers Selected at Random

- How do you access the documents that are critical to your
  job?

- How do you know you have the latest version?

Example Auditor Questions.   Example questions asked at
the relevant levels of the organization with respect to document control are presented in Table 5-10.

Sources of Objective Evidence to Validate Conformance.
Objective evidence that shows the requirements of this element have been met will be the controlled documents themselves. For ease of auditing, the organization might want to
develop a matrix that specifies which documents are controlled
under the document control procedure and where they are
located. The following information should be available for all
controlled documents: name of owner or coordinator, name of
approver, date of last update, and next scheduled review date.

Controlling obsolete documents is more difficult for hard-
copy documents than electronic documents. Electronic documents can be controlled by referencing the latest version as
that which is on-line, with all hard-copy printouts being
uncontrolled documents. Hard-copy documents can be managed by maintaining a list of who has the controlled documents and ensuring that the outdated copies are collected for
retention or disposal at the time the updated copy is issued.

**Examples of Core Elements of the EMS
and Related Documentation**

Core Elements

✓ Environmental policy
✓ Documented environmental management programs
✓ Documented roles and responsibilities
✓ Procedures required by ISO 14001

Related Documentation

✓ EMS audit methodology
✓ EMS awareness training materials/documents
✓ Procedures related to the operation of the wastewater
  treatment facility
✓ Procedures related to the operation of the chemical distribution
  and waste management center
✓ Chemical authorization procedure
✓ Hazardous waste handling procedure
✓ Emergency response plan

FIGURE 5-11.   Examples of core elements of the EMS
and related documentation.

TABLE 5-10. Example Auditor Questions Pertaining to the Organization's EMS Document Control

---

Questions Asked of EMS Management Representative and/or Environmental Staff

- Show me your procedure for document control. (Note: This procedure will be tested.)

- Who has authority to establish, approve, and issue documents?

- How do you control hard-copy documents?

- How do you control electronic documents?

- How do you back up electronic versions of controlled documents?

- How do you ensure that obsolete documents are removed from the system?

- What are the security controls for your electronic documents?

- How do you determine when revisions to documents are needed?

- How do you communicate revisions?

- What are the security controls for you electronic documents?

- What forms or checklists do you have that are controlled documents?

Questions Asked of Employees and Managers Selected at Random

- What operating procedures do you have?

- How are these procedure controlled?

- When were they last reviewed or updated?

- How do you know you have the latest version?

- Where is the master copy of the controlled document?

- Who can change the master copy and how is the master copy of the controlled document changed?

OPERATIONAL CONTROL (SECTION 4.4.6)

Operational control of the organization's activities, products, and services to ensure conformance to the environmental policy and overall EMS is critical. Typically, testing conformance to this element of the standard requires extensive interviewing with employees performing "operations" jobs.

Audit Methodology.   Determine if the organization has

- Identified operations and activities that are associated with the identified significant aspects in line with its policy, objectives, and targets
- Established and maintained documented procedures to cover situations where their absence could lead to deviations from the environmental policy—including the commitment to comply with relevant environmental legislation and regulations—and the objectives and targets
- Stipulated operating criteria in the procedures
- Established and maintained procedures related to the identifiable significant environmental aspects of goods and services used by the organization
- Communicated relevant procedures and requirements to suppliers and contractors

Test that

- Procedures are implemented according to planned arrangements
- On-site contractors and suppliers are aware of appropriate operational procedures and are implementing them according to planned arrangements

Example Auditor Questions.   Example questions asked at the relevant levels of the organization with respect to document control are presented in Table 5-11.

Sources of Objective Evidence to Validate Conformance. Although not required by the standard, it is useful to develop a list or otherwise identify procedures that are necessary to

Table 5-11. Example Auditor Questions Pertaining to the Organization's Operational Control

Questions Asked of the EMS Management Representative and/or Environmental Staff

- What are the activities associated with your significant environmental aspects?
- What are the procedures necessary to maintain operation control of the EMS?
- What procedures are necessary for maintaining control of the following: wastewater treatment, waste management, energy management, spill response, chemical control, control of air and water emissions?
- What is the process for maintaining operational control of suppliers as it relates to your EMS?
- How are relevant environmental procedures communicated to suppliers and contractors?
- How are operational control procedures identified within the core and/or related documents within the EMS?

Questions Asked of Employees and Managers Selected at Random

- What procedures do you have to ensure that legal requirements are met?
- What forms and/or checklists do you use with respect to elements of the EMS? Are these forms and/or checklists identified as an environmental record?
- Who has the master controlled copy of the form and/or checklist?
- Show me that the procedures specify how to manage chemicals and waste. (Implementation of the procedure will then be tested.)
- What do you do if you think changes to the procedures are needed?
- How do you know in which bay to put which chemicals?
- Show me that this chemical (found in the area) is authorized.
- Show me the Material Safety Data Sheet (MSDS) for this chemical (found in the area).

TABLE **5-11.** Example Auditor Questions Pertaining to the
Organization's Operational Control (*Continued*)

- How do you know that all chemicals that come onto the site are authorized?
- At what settings (for water or chemical flow rate) is this machine supposed to be operated?
- Who checks to see if the settings are correct and how often are they checked?
- What do you do if the high-level alarm goes off?
- Where does this drain go? What is allowed to be discharged to it?
- How have you been trained to mix these chemicals?

maintain operational control of EMS. Essentially, the main source of objective evidence for validating conformance to operational control will be the procedures themselves and verification that they are being implemented as planned. Typically, this section of the audit requires extensive employee interviews and perhaps demonstrations of operations.

EMERGENCY PREPAREDNESS AND RESPONSE (SECTION 4.4.7)

Audit Methodology. Determine if the organization has established, maintained, and tested (where practicable) procedures to identify potential for and to respond to accidents and emergency situations and for preventing and mitigating the environmental impacts that may be associated with them. Validate that

- The organization follows the established emergency preparedness and response procedures
- The location has reviewed and revised, where necessary, the emergency preparedness, and response procedures, in particular after the occurrence of accidents or emergency situations

Example Auditor Questions.   Example questions asked at the relevant levels of the organization with respect to emergency preparedness and response are presented in Table 5-12.

Sources of Objective Evidence to Validate Conformance. The main document that needs to be available to show conformance to this section of ISO 14001 is the emergency preparedness and response procedure. A summary of what elements are typically included in one of these plans is presented in Figure 5-12. In addition to this procedure, there are other sources of objective evidence that the organization might want to have available. These include training records for emergency response personnel, copies of the organization's last inspections of emergency equipment, and documentation showing the last test of the procedures and the dates of the last review or revision of the emergency procedures.

CHECKING AND CORRECTIVE ACTION (SECTION 4.5)

MONITORING AND MEASUREMENT (SECTION 4.5.1)

Audit Methodology.   Determine if

- The organization has established and maintained procedures to monitor and measure, on a regular basis, the key characteristics of its operations and activities that can have a significant impact on the environment
- Monitoring equipment is calibrated according to schedule and records are maintained
- The organization has established and implemented a procedure to periodically evaluate compliance with relevant environmental legislation and regulations

Validate that there are the procedure(s) for monitoring and measuring key characteristics on a regular basis and that the procedure(s) includes

TABLE **5-12.** Example Auditor Questions Pertaining to the Organization's Emergency Preparedness and Response

Questions Asked of the EMS Management Representative and/or Emergency Preparedness and Response Manager

- Show me your emergency response procedures.
- How did you identify areas and/or activities or services with potential environmental emergencies?
- What are these areas and/or activities or services?
- How often do you test your emergency response procedures?
- Show me a copy of your last emergency test.
- Describe the results of the last time you tested them.
- As a result of the test, did you make any changes?
- What is the procedure for making changes to these procedures?
- How do you record incidents?
- Show me a copy of your last incident report.
- If there was an emergency that could affect your neighbors, how would you handle it?
- What is the role of security department during emergencies?
- How do you train emergency response personnel?

Questions Asked of Employees and Managers Selected at Random (Security and Others Who Implement the Emergency Plan Will Be Questioned in Detail on This Element.)

- If there was an emergency, what would you do?
- Have you taken any training with respect to emergency response?
- When was the last emergency in which you were involved?
- Do you know what corrective action was taken after the incident?

- Method(s) to record information to track performance
- The identification of relevant operational controls
- Method(s) to track conformance with the location's environmental objectives and targets

---

**Elements that Might be Included in an Emergency
Preparedness and Response Manual Related to the EMS**

✓ Roles, responsibilities, and authorities for responding to chemical
  emergencies
✓ Emergency notification procedure, including phone numbers for
  those who need to be contacted
✓ Emergency response procedure, including a description of
  mitigation techniques
✓ Description of required training for emergency responders
✓ Communication procedure for incidents that may have off-site
  impacts beyond the location boundaries
✓ List of emergency equipment and location of such
✓ Procedure for conducting root cause analysis
✓ Procedure for tracking corrective and preventive action
✓ Reporting (legal/regulatory) requirements
✓ List of suppliers and/or contractors that may be required during
  emergency response situations

FIGURE 5-12. Elements that might be included in an emergency preparedness and response manual related to the EMS.

Test that

- Data has been gathered according to procedure
- The organization has adhered its schedule for calibration of its monitoring equipment
- The organization has implemented its procedure for evaluating compliance with relevant environmental legislation, regulations, and permits.

Example Auditor Questions.   Example questions asked at the relevant levels of the organization with respect to monitoring and measurement are presented in Table 5-13.

Sources of Objective Evidence to Validate Conformance. Although not required specifically by the standard, it is useful for an organization to develop a master list of key characteristics that are monitored and/or measured, Likewise, it is useful to have a master list of the equipment that should be calibrated as part of the EMS. (Note: In some cases, the equipment may be tested during scheduled preventive maintenance instead of calibrated, per se; examples are high-level alarms and oil burner efficiency.) These lists may be maintained by a centralized department or by a single individual. Examples of what an organization might consider key characteristics and some potential measurement methods are presented in Table 5-14. Examples of equipment that might need calibration as part of the EMS, along with calibration frequency, are presented in Table 5-15.

NONCONFORMANCE AND PREVENTIVE AND CORRECTIVE ACTION (SECTION 4.5.2)

Audit Methodology.   Determine if the organization has established and implemented a procedure for handling nonconformance and for corrective and preventive action. Validate that

- A procedure exists for handling nonconformance and corrective and preventive action

TABLE 5-13. Example Auditor Questions Pertaining to the Organization's Monitoring and Measurement Process

---

Questions Asked of the EMS Management Representative and/or Environmental Staff

- What are the key characteristics that you monitor and measure?

- Show me evidence of the monitoring data.

- How do you track performance with respect to objectives and targets?

- What monitoring and measuring equipment is associated with key characteristics of the EMS?

- Where is the calibration schedule for this equipment?

- Show me your procedures for periodically evaluating compliance with relevant environmental legislation and regulations.

- Show me evidence that you have implemented these procedures.

- Who performs preventive maintenance of environmental equipment?

- Show me the preventive maintenance schedule.

Questions Asked of Employees and Managers Selected at Random

- What monitoring and measurement is performed by this department?

- What pieces of equipment need to be calibrated?

- Show me the calibration procedure and schedule.

- Show me your calibration records.

- What pieces of equipment are calibrated by contractors or venders (i.e., scales and electricity and water meters)? Show me evidence that the calibration is being performed.

- Do you use off-site laboratories for sample analysis? If so, how do you verify their equipment calibration?

TABLE **5-14.** Examples of Key Characteristics and Measurement Methods

Air Emissions
Key Characteristics

- Parameters listed in organization's air emissions permit
- Other parameters that the organization discharges into the air through abated and unabated stacks that might cause environmental impact, including ozone precursors, greenhouse gases, and toxics
- Flammable gases
- Chemical mass balance
- Fuel consumption
- Boiler efficiency

Measurement Methods

- On-line measurement systems such as a flame ionization detector (FID), photo-ionization detector (PID), combustible gas indicator (CGI), on-line mass spectrometer (MS), or other appropriate method
- Periodic stack sampling using a sampling train and an appropriate collection and analytical method
- Chemical feed meters
- Preventive maintenance tests

Chemical Use
Key Characteristic

- Quantity of chemicals consumed

Measurement Methods

- Trends of use, as related to activity index, based on purchasing records per time period
- Trends of use, as related to activity index, from on-line bulk chemical monitoring system per time period
- Trends of use from other usage monitoring, based on activity index, per time period

TABLE 5-14. Examples of Key Characteristics and Measurement Methods (*Continued*)

Energy Use
Key Characteristic

- Energy consumed

Measurement Methods

- Electric bills, based on activity index, per time period
- Periodic audits of manufacturing areas, on weekends and other times of nonproduction, to ensure equipment is turned off when not in use
- Periodic audits of office areas, on weekends and evening hours, to ensure lights and computers are turned off when not in use
- Direct meter readings

Recycling Activities
Key Characteristic

- Materials and wastes that have recycle potential

Measurement Methods

- Weight or volume of material or waste recycled, as related to activity index, per time period
- Revenue generated from recycling activities, as related to activity index, per time period
- Disposal costs avoidance

Stormwater Discharges
Key Characteristics

- Chemicals that might be found in stormwater as a result of the organization's activities
- Parameters that are listed on the organization's wastewater discharge permit
- Oil and grease from parking lot runoff
- Other parameters that might be found in stormwater as a result of the organization's activities
- Rate of runoff from site

TABLE **5-14.** Examples of Key Characteristics and Measurement Methods (*Continued*)

Measurement Methods

- Grab sampling on a periodic basis (i.e., monthly or whenever it rains) at selected outfalls; analysis of specified "indicator" parameters such as pH, chemical oxygen demand (COD), and oil and grease
- Composite sampling during first hour of rainfall event on a semiannual basis, if required under stormwater permit; analysis of publicly reported chemicals and other specified parameters
- Flow rate sampling during all or specified sampling events
- Stormwater runoff volume

Treatment Efficiencies
Key Characteristics

- Treatment efficiencies specified in organization's permits such as efficiencies in waste water treatment, air abatement, incineration, and other prescribed technology efficiencies
- Treatment efficiencies required for state or country reports and/or other public reports
- Other treatment efficiencies considered important by the organization such as water purification efficiencies, chemical reprocessing efficiencies, and other efficiencies

Measurement Methods

- Mass balance
- Analytical comparisons of influents and effluents
- Use of efficiency factors
- Manufacturer's literature estimates

Unplanned Releases
Key Characteristics

- Unplanned emissions or effluents exceeding permit values
- Chemical spills that breach secondary containment
- Other chemical spills
- Large-volume water spills, such as process water, chilled water, and treated wastewater spills

Table 5-14. Examples of Key Characteristics and Measurement
Methods (*Continued*)

Measurement Methods

- Mechanisms for unplanned release reporting and tracking
- Trends of unplanned releases per time period

Waste Generation
Key Characteristics

- Hazardous waste generation, including ignitable, corrosive, reactive, toxic waste, and that listed by a regulatory agency as hazardous
- Nonhazardous waste generation, including wood pallets, paper, and aluminum cans

Measurement Methods

- Amount generated, as related to activity index, per time period (i.e., monthly, quarterly, semiannually, or annually)
- Disposal and/or treatment costs, as related to activity index, per time period
- Number of waste shipments per time period

Wastewater Discharges
Key Characteristics

- Parameters in discharge permit
- Parameters in state, country, or city regulations
- Other parameters that could impact the environment, as applicable to area, such as nutrients, metals, toxics, and other parameters
- Wastewater volumes

Measurement Methods

- Periodic grab or composite sampling of outfalls for specified parameters, per permit requirements
- On-line monitoring of certain parameters such as pH
- Additional daily, weekly, or monthly grab or composite sampling, using appropriate sampling method
- Screening sampling for batch treatment or process observation
- Flow meters

TABLE **5-14.** Examples of Key Characteristics and Measurement
Methods (*Continued*)

Water Use
Key Characteristics

- Amount of water consumed
- Amount of purified water consumed, such as water treated by filtration, reverse osmosis, or deionization
- Amount of irrigation water consumed

Measurement Methods

- Trends of use, as related to activity index, based on water meters per time period
- Trends of use, as related to activity index, based on water bills per time period

- The procedure defines responsibility and authority for handling and investigating nonconformances
- Nonconformances are handled according to the procedure and that proper corrective action has been taken based on the magnitude of the incident

Example Auditor Questions.   Example questions asked at the relevant levels of the organization with respect to nonconformance and corrective and preventive action are presented in Table 5-16.

Sources of Objective Evidence to Validate Conformance. Sources of objective evidence to demonstrate conformance to this element of ISO 14001 might include (1) a completed nonconformance form, (2) a nonconformance tracking log, and (3) physical evidence that the corrective and preventive actions have been implemented. A point to remember about this element of the standard is that the auditor is looking for preventive action and not just corrective action.

RECORDS (SECTION 4.5.3)

Audit Methodology.   Determine if the organization has established, implemented, and maintained a procedure(s) for

TABLE 5-15. Equipment that Might Need Calibration within the EMS

EQUIPMENT	FREQUENCY
Air Monitoring Equipment	
On-line flame ionization detector	Weekly (self-calibration)
On-line mass spectrometer	Weekly (self-calibration)
On-line gas detector	Daily (self-calibration)
On-line carbon monoxide detector	Daily (self-calibration)
Flow meter	Monthly
High concentration alarm(s)	Monthly
Bulk Chemical Monitoring Equipment	
Tank level gauge	Monthly
Flow meter	Monthly
Tank overflow alarm	Quarterly
Leak detection system	Quarterly
Temperature sensor	Quarterly
Pressure sensor	Quarterly
Scales	Annually
Energy Monitoring Equipment	
Power use meter	Annually
Run time meter	Annually
Incident Management Equipment	
Oxygen/lower explosive limit meter	Before each use
pH sensor	Before each use
Leak detection sensors	Quarterly
Process Monitoring Equipment	
Chemical metering sensor	Weekly
Water meter	Weekly
Part counter	Monthly
Temperature sensor	Quarterly
Pressure sensor	Quarterly

TABLE **5-15.** Equipment that Might Need Calibration within the EMS (*Continued*)

EQUIPMENT	FREQUENCY
Stormwater Monitoring Equipment	
pH sensor	Weekly
Flow recorder	Quarterly
Wastewater Monitoring Equipment	
Atomic absorption	Daily
Gas chromatograph	Daily
Mass spectrometer	Daily
pH sensor	Daily
Colorimetric spectrophotometer	Daily
Flow meter	Monthly
High-level alarms	Quarterly
Waste/Recyclable Materials Management Equipment	
Building leak detection sensors	Quarterly
Building lower explosive limit meters	Quarterly
Scales	Annually
Water Usage Monitoring Equipment	
Process flow meter(s)	Quarterly
Cooling tower make-up flow meter(s)	Quarterly
Cooling tower blow-down flow meter(s)	Quarterly
Irrigation system meter(s)	Annually
City inlet flow meter(s)	Annually

the identification, maintenance, and disposition of environmental records. Validate that the procedure(s) is being followed. Test records to validate that they have the following attributes:

- They are legible.
- They are identifiable and traceable to the activity, product, or service involved.

TABLE 5-16. Example Auditor Questions Pertaining to the Organization's Handling and Investigating of Nonconformance and for Taking Preventive and Corrective Action.

Questions Asked of the EMS Management Representative and/or Environmental Staff

- How do you identify nonconformance to your EMS?
- How do you handle nonconformances?
- How do you evaluate root cause of the nonconformance?
- Do you maintain a log? Show me.
- Who is responsible to put a corrective and preventative action plan in place and track to closure?
- How is this plan administered?
- Do you consider it a nonconformance if you don't meet your objectives and targets?
- Who is responsible to verify closure of nonconformance?

Questions Asked of Employees and Managers Selected at Random (Questions Typically Will Be Directed to Managers Who Have Had a Nonconformance Identified.)

- What happens when there is a nonconformance identified in your area?
- How do you make sure it doesn't happen again?
- Show me your corrective action for this nonconformance identified during the last EMS audit.
- How do you determine preventive action?

- They are stored and maintained in such a way that they are readily retrievable and protected against damage, deterioration, and loss.
- Their retention times are established and recorded and record retention conforms.

Example Auditor Questions.   Example questions asked at the relevant levels of the organization with respect to environmental records are presented in Table 5-17.

Sources of Objective Evidence to Validate Conformance. Environmental records themselves are the source of objective evidence to validate conformance to the requirements of this element of ISO 14001. In addition, although not required by the standard, the organization might want to maintain a master list of environmental records and their location. An example master list is presented in Figure 15-13. At a minimum, records kept should include training records, calibration records, reports to governmental bodies, and permits.

EMS AUDIT (SECTION 4.5.4)

Audit Methodology. Determine if the organization has established and implemented EMS audit procedure(s) and program(s) and has reported audit results to management. Validate that

- The organization has an audit procedure and program that includes the following:

TABLE 5-17. Example Auditor Questions Pertaining to the Organization's Environmental Records.

Questions Asked of the EMS Management Representative and/or Environmental Staff

- What is your procedure to identify and maintain environmental records?
- What records are considered environmental records?
- Where are these records stored?
- How is damage prevented?
- Who determines retention times?
- How do you dispose of environmental records?

Questions Asked of Employees and Managers Selected at Random Who have Environmental Records

- What environmental records do you keep?
- How long do you retain these records?
- What do you do with the records when their retention time has expired?

XYZ List of Environmental Records

Environmental Record	Retention Time	Location
List of Significant Environmental Aspects	three years	A
Documented Objectives and Targets	current version	A
Calibration records for pH meter wastewater treatment facility	two years	B
Calibration records for pH meter environmental laboratory	one year	B
Calibration records for truck scale at chemical building	three years	B
Employee Training for hazardous waste management	twenty five years	C
Employee Training for wastewater treatment operation	twenty five years	C
Monthly solid waste generation volumes	three years	CDC
Wastewater monthly report	five years	B
Top management review	five years	A
Communication to external interested parties	ten years	D
Results of emergency preparedness testing	three years	E
EMS audit reports	three years	A
Nonconformity / corrective action report	three years	A
Hazardous Waste Warehouse inspection checklist	three years	CDC
Hazardous Waste Shipping Documents	twenty five years	CDC
Incident reports	ten years	E
Results of compliance reviews	ten years	B
Environmental Laboratory Analysis Reports	ten years	B

Location Key:

A -- Environmental Management Representative's Office
B -- Environmental Programs File Room
C -- Human Resources File Room
CDC -- Chemical Distribution Center
D -- Communication Manager's Office
E -- Security Central Office

FIGURE 5-13.   Example of an environmental records master list.

Audit scope

Audit frequency and methodology

Responsibilities and requirements for conducting the audits and reporting results

- The organization has conducted EMS audits according to its procedure(s) and program(s) and has reported the results to management

- EMS audits are conducted by auditors who can act impartially and objectively about the area being audited

- EMS audits are conducted by auditors with adequate experience and/or training

- EMS audits are conducted according to the documented schedule

Test that nonconformances identified by the EMS audit have been corrected.

Example Auditor Questions.   Example questions asked at the relevant levels of the organization with respect to EMS audits are presented in Table 5-18.

Sources of Objective Evidence to Validate Conformance. The auditors will be requesting the following as sources of objective evidence that the organization conforms to the requirements of this section of ISO 14001: a documented audit plan and schedule, training and competence requirements for the auditor, results of last EMS audit available (this is a record), and audit methodology. The organization's process for evaluating compliance with legal requirements is not part of the EMS audit process and should be used to provide objective evidence for Section 4.5.1— monitoring and measurement—only.

MANAGEMENT REVIEW (SECTION 4.6)

AUDIT METHODOLOGY

Determine if top management periodically reviews the EMS to determine its continuing suitability, adequacy, and effective-

TABLE 5-18. Example Auditor Questions Pertaining to the Organization's EMS Audits

---

Questions Asked of the EMS Management Representative and/or Environmental Staff

- How did you determine an audit schedule?
- What was the scope of the last EMS audit?
- What is your audit methodology?
- How do you select the auditors?
- How do you determine auditor qualifications and required training?
- Who can raise a nonconformance?
- Who gets the audit reports?
- How have you defined responsibility and authority to handle nonconformance found from the EMS audit?
- Who is responsible for making sure audit nonconformances are corrected?
- What were the results of your last EMS audit?
- How did you consider results of previous audits in preparing the audit schedule?

Questions Asked of Employees and Managers Selected at Random

- Have you had an EMS audit in your area?
- What were the results?
- Who is responsible for putting a plan in place to close non-conformances?
- Who is responsible for approving corrective and preventive actions?
- How do you verify closure of the nonconformances?

ness. Validate that management reviews occur and that the adequate data is presented for management to determine the suitability, adequacy, and effectiveness of the EMS. Test that the review(s) covered the following:

- The need for changes to the policy, objectives, and EMS elements
- Environmental performance
- Changes to activities, products, and services that have resulted in amendments or other changes to the EMS
- The results of EMS audits
- Opportunities for continual improvement

Example Auditor Questions

Example questions asked at the relevant levels of the organization with respect to management review are presented in Table 5-19.

Sources of Objective Evidence to Validate Conformance

For objective evidence that the requirements of this element of ISO 14001 have been met, the organization (in particular, the management representative for the EMS) will want to keep appropriate documentation that the review took place. Typically, this documentation will include the date of the review, a list of attendees, a copy of the agenda, presentation materials, and meeting minutes. These meeting minutes should include questions and recommendations made by top management for continual improvement as well as top management's evaluation of the suitability, adequacy, and effectiveness of the EMS.

TABLE 5-19. Example Auditor Questions Pertaining to the Organization's Management Review

---

Questions Asked of EMS Management Representative

- Who is top management?

- How do you collect the information needed for management review?

- How often do you have management review?

- Did you discuss the policy and need for changes to the policy?

- Is the review documented?

- What was date of last review?

- Who attended?

- What elements were reviewed?

- Were there any recommendations for continual improvement?

- Will they be incorporated in objectives and targets?

- As a result of the management review, were there any changes to the EMS?

- Can you show me the record of the last top management review?

Questions Asked of Top Management

- Have there been changes to your business that have required an update to your environmental policy or EMS?

- How often are you updated with respect to achieving objectives and targets?

- What do you do if you are not meeting your objectives and targets?

- How do you determine if your EMS is suitable, effective, adequate?

- How often do you have a management review of the EMS?

- When are the results of the EMS audit presented to you?

REGISTRATION AUDITS

Once your organization has aligned its environmental management system (EMS) with the requirements of ISO 14001, the next thing to consider is registration. Reasons to register will vary from organization to organization but might include things such as customer or stakeholder requirements, marketplace advantage, validation that the organization is executing per planned arrangements, evaluation of the efficiency and effectiveness of the EMS, and identification of opportunities for continual improvement of the EMS. Whatever the reason the organization chooses to register, the registrar selected to conduct the registration audit should be one who can add value to your EMS.

THE REGISTRATION AUDIT PROCESS

As mentioned in Chapter 2, an EMS audit—whether it is an internal EMS audit or a registration audit—is a systematic, objective, and unbiased verification process to determine whether or not an organization conforms to planned arrangement for environmental management. For the case of registration, the audit must also be a third-party audit. Once top management has made the decision to proceed with registration, it is important to understand the registration process.

Registration audits are generally conducted as a two-step process, which includes an initial audit and a main or final audit. In addition, an organization may elect to conduct an optional preassessment audit. The specifics of each type of registration audit may vary slightly from registrar to registrar, but they basically include the elements defined below.

PREASSESSMENT AUDIT

A preassessment audit is optional and is not part of the registration process. During this audit the registrar basically conducts a "mini" combined initial and main audit. There will be some preaudit planning and an agenda, opening meeting, site tour, document review, employee interviews, closing meeting, and nonconformity reports. A good preassessment audit will test elements of the entire system.

Because the preassessment audit is not formally part of the audit process, the organization is not required to take any corrective and preventive action for those nonconformances identified, although most organizations will, of course, want to correct what was found. In addition, registrars are expected to "forget" everything they read, saw, and heard once they leave a location. Because it is truly not part of the formal audit process, when they return for initial or main audits, preconceived notions of how the EMS has been implemented and maintained are not allowed. Thus, a preassessment offers a unique opportunity for employees to experience the audit process before the actual registration audit begins.

INITIAL AUDIT

During the initial audit the registrar will come to the location, conduct a document review, and assess if all the elements of the ISO 14001 standard have been addressed. Generally, there will be very limited employee involvement other than the immediate EMS management representative and the environmental staff. The intent is to verify that the system conforms with the intent of the standard. Typically, there is an emphasis on the environmental policy and the organization's support of its elements, identification of significant aspects, the setting of objectives and targets, and top management review. Nonconformances identified during this audit must be closed within 90 days and/or before the start of the final or main audit.

FINAL OR MAIN AUDIT

During the final or main audit, the registrar verifies that the entire system is implemented and maintained. In essence, the entire system will be tested. The audit will include a review of some of the same documentation that was covered during the initial audit and any corrective and preventive actions resulting from nonconformances found at that time. In addition, there will be an in-depth review of all the requirements of ISO 14001, particularly environmental management programs, operational control, monitoring and measurement, EMS documentation, and environmental records. Throughout this audit, there will be extensive employee interviews.

As with the initial audit, nonconformances identified must be cleared with the registrar within 90 days. Once cleared, the lead auditor will recommend the organization for registration. At this point, an independent person within the registrar's head office will review all information provided by the lead auditor, including audit scope, auditor notes, nonconformances identified, and closure of these nonconformances. Once the review proves that all aspects of the audit were satisfactory, the registrar will issue a certificate of registration to the standard. This will include the name of the organization, name of registrar, operational scope, date of issue and accreditation body (i.e., U.S. Registrar Accreditation Board, United Kingdom Accreditation System, Japan Accreditation Board).

MAINTENANCE OF REGISTRATION

Once the organization is registered to ISO 14001, it will have to demonstrate periodically that the system continues to be implemented, maintained, and continually improved. This is carried out through *surveillance audits*. Frequency of these visits will depend on the registrar and the accreditation bodies on the certificate, although, typically, surveillance audits are conducted on a 6-month or 1-year cycle. If the organization elects to participate in 6-month surveillance audits, typically all the elements are not tested each time—although some key elements such as legal and other requirements, objectives and targets and the environmental management program, and nonconformance and corrective and preventive action may be. Over a 3-year period, however, all elements are tested. In addition, a complete EMS audit may be required every 3 years.

SELECTING A REGISTRAR

The registrar has been invited to help the organization ensure its EMS meets all requirements of ISO 14001 and planned arrangements. There is teamwork and partnering between the registrar and the organization; thus, it is extremely important to select a registrar that meets as many of the organization's needs as possible. There are several primary factors to consider when selecting a registrar, including the following:

- *Experience within the industry* The registrar should be familiar with the specifics of the organization's industry, including legal requirements, knowledge of processes, and general categories of chemicals used and waste generated. The organization should develop the scope of the registration early so that potential registrars can provide detailed information about their qualifications with respect to the needs of the organization and scope of the registration.

- *Accreditation* The organization will need to verify that the registrar is accredited by a registrar accreditation entity. Because there are numerous accreditation bodies around the world, the organization should determine which entity best suits its business needs.

- *Registration experience* Many registrars have the majority of their experience with ISO 9000 and have simply "added on" ISO 14001 registrations. The organization will want to know specifics about the registrar's experience with ISO 14001 and, in some cases, the European Union's Environmental Management Audit Scheme (EMAS).

- *References*   The organization should request and contact references. Information about how the registrar's performance should be requested. Questions to be answered include: Does the registrar respond to inquiries in a timely fashion? Are audits conducted professionally? Do the auditors communicate effectively? Can they conduct the audit in a objective and impartial manner?

- *Worldwide coverage*   Depending on the number and location of facilities that are to be registered, the organization might find it necessary to select a registrar with international presence. Many international companies will want to require the registrar to be familiar with the national legal requirements, language, and culture. In addition, the registrar should have enough auditors to provide adequate coverage and have them in close proximity to the company's facilities to reduce costs associated with travel.

- *Auditor qualifications*   The organization will want to understand the auditor(s) qualifications, including education, technical experience, and experience in auditing an EMS. In addition, the organization will want to identify and understand the registrar's policy on using contractors or consultants as auditors. (Note: In many situations contractors are retained to provide the necessary expertise for a specific industry.)

- *Business structure*   The organization should inquire about the business structure of the registrar with questions such as who is the business focal point for scheduling the registration audit(s) and handling administrative issues? If there is a technical issue or disagreement over an interpretation issues, who has authority to resolve them? How does it address noncompliance issues?

- *Audit costs*   Although not the sole criteria for selecting a registrar, the organization will want to compare cost of conducting the audits. In addition, the organization will want to look for cost efficiencies such as travel logistics and combining audit events when possible.

In addition to the above considerations, the organization must realize that the registrar will have access to the details of its business and operational practices. Thus, it is important to verify that the registrar maintains its confidentiality agreements. Finally, the registrar must be able to add value to the organization's EMS.

In some cases, secondary factors may come into play when considering the selection of the registrar. These include

- *Registration to ISO 9000*   Does the registrar have a quality management system that is registered to ISO 9000? If the registrar has implemented a quality management system, this will lend some assurance that its top management is involved with the planning and implementation of the registration process, with the result being (it is hoped) a consistent registration system.

- *Ability to combine ISO audits*   The organization might want to ask if the registrar is capable of conducting ISO 14001 and 9001/2 audits during the same audit session. In some cases, this may improve efficiency of the system audits and minimize disturbance in the workplace.

- *Ability to work with other registrars*   Depending on the structure of the organization's EMS and past experience with ISO 9000 audits, it might be preferable to have two different registrars for ISO 9001/2 and ISO 14001. If this is the case and the organization wants the audits conducted during the same session, both registrars must have the ability and willingness to work with one another. For efficiency, it may be advantageous to have a single, coordinated team of auditors from the two registrars. Should this occur, the organization will want to make sure that all auditor roles and responsibilities are clearly understood.

PREPARING FOR THE REGISTRATION AUDIT

The importance of preaudit preparation cannot be overstated. Below are some hints for ensuring that the registration audit goes as smoothly as possible:

- Ensure that all elements of ISO 14001 have been implemented before beginning the registration audit. In particular, at least one complete EMS audit and a top management review needs to be completed before the registration audit can take place.

- Make it clear ahead of time to all involved in the audit process that the ISO 14001 registration audit is a "friendly" audit. Many managers and personnel get nervous when being audited, so understanding that the purpose of the audit is to improve the EMS, not pinpoint blame for inadequacies can put those being audited at ease.

- Ensure that employees and contractors know that they will be interviewed as part of the audit. This may be a new experience for many employees. Ensure they understand that they will be asked about their roles and responsibilities within the EMS. Make it clear that this is not a time for them to vent frustration by criticizing others or pointing out the shortcomings of the EMS.

- Ensure that employees and contractors understand the terms the auditors will be using. Some examples include

Objective evidence   Objective evidence is a means of verifying that the organization is doing what it said it would do. Examples of objective evidence may include things such as a documented procedure that is being followed, meeting minutes from a management review, a record of instrument calibration, the organization's documented environmental policy, a list of significant aspects, and other evidence.

Environmental record   An environmental record is evidence that something has occurred. There are numerous types of environmental records, such as meeting minutes, calibration records, audit reports, data summaries, discharge reports, letters to interested parties, completed checklists, organization charts, and others.

Controlled document   A controlled document is one that is maintained per an established document control procedure. This includes identification of author, approver, and latest version.

In addition to these terms, employees and contractors need to be familiar with the organization's environmental policy, significant aspects, and objectives and targets.

- If the auditors are to tour the activities, products, and services of the location, plan the tour route carefully and inform managers of the approximate time the tour group will reach their areas. Walk the route in advance to understand what personal protective equipment is required, the time needed to get from one area to another, and other such logistics.

- Ensure that good housekeeping practices are maintained in all areas to be audited. A well-kept area provides the auditors with a first impression of a well-maintained EMS.

- Conduct "dry runs" of questions the auditors will ask and answers that they will expect to hear. Include top management, line managers, environmental staff, legal and communication personnel, employees, and contractors in this process.

- Review key documents and procedures and verify that employees are following them.

- Keep top management briefed of progress toward registration readiness.

SELF-DECLARATION

Typically, an organization will use an accredited registrar to conduct its ISO 14001 audits. There is another option for the organization, however, and that is self-declaration. Essentially, the main benefit to self-declaration is that it allows organizations (typically smaller ones) to state publicly their environmental commitment—through conformance to the requirements of ISO 14001—without having the expense of a registration audit. The drawback, of course, is that self-declaration does not have the credence with the public that a third-party registration has. Organizations should weigh carefully the pros and cons when deciding to self-declare conformance to the standard or validate conformance through a registration audit.

PUTTING
IT ALL
TOGETHER

THE VALUE OF ISO 14001

BENEFITS FROM IMPLEMENTING ISO 14001

Once an organization has implemented ISO 14001, what benefits can be expected? Examples of these, including general benefits and benefits per ISO 14001 element, are presented in Table 7-1.

BENEFITS FROM REGISTRATION TO ISO 14001

As seen in Table 7-1, the benefits from implementing ISO 14001 are numerous. Registration to the standard also offers an organization benefits. These are summarized in Table 7-2.

CONTINUAL IMPROVEMENT ACTIVITIES

A very important element of ISO 14001 is the requirement for the organization to continually improve its EMS. Examples of types of continual improvements that the organization might strive for are presented in Figure 7-1.

TABLE 7-1. Benefits that an Organization Can Expect to Derive from the Implementation of ISO 14001

General

- If an organization chooses to develop an environmental management system (EMS) manual that aligns its system with ISO 14001 (and the authors suggest doing so), the structure, roles, and responsibilities for the EMS—element by element—are clearly defined.

- EMS procedures are in place and documented, as required by the standard, and thus the EMS is system and not person dependent.

- Environmental management is considered the responsibility all managers and employees, and not solely the environmental staff.

- The documented EMS includes roles, responsibilities, and interaction with the EMS of formerly disconnected functions such as procurement, facilities maintenance, and distribution.

- ISO 14001 provides the organization with a framework that can add efficiency and effectiveness to the EMS.

- The organization can prioritize environmental projects by setting objectives and targets, which can lead to proactive, cost-efficient environmental management.

Environmental Policy

- Top management commits to comply with legal requirements, prevention of pollution, and continual improvement of the EMS.

- All employees and contractors are made aware of the organization's environmental policy and ways that they can support the policy objectives.

- Methods of supporting the policy are considered on an ongoing basis.

Environmental Aspects

- A thorough review of environmental aspects and impacts is performed; this ensures that inputs and outputs of activities, products, and services are taken into account when determining the organization's significant aspects.

- Procedures are set up to identify significant aspects.

Table 7-1. Benefits that an Organization Can Expect to Derive from the Implementation of ISO 14001 (*Continued*)

Objectives and Targets

- The organization sets and tracks formal objectives and targets for environmental improvements to the EMS.

- These objectives and targets are monitored and measured closely and reported to top management on a periodic basis.

Environmental Management Program

- The organization has a documented program for applicable objectives and targets. These programs include the responsible person, time frame, and means for achieving the objective and target; this makes the execution of the plan system dependent.

- Through the EMS audit and registration process (if applicable), some inconsistencies in reporting measurements toward meeting objectives and targets may be identified and corrected.

Structure and Responsibility

- The EMS clearly defines who has authority and responsibility within the EMS; cross-functional responsibilities are delineated, as are the responsibilities of every employee. This information is communicated through the training and awareness programs.

Training, Awareness, and Competence

- Training needs for employees and contractors—which may not have been formally defined—are identified.

- Training records are kept.

- Competence through training, education, and/or experience for those employees whose job could potentially impact the environment is required.

Communications

- Communications to employees about the policy, objectives and targets, significant aspects, and other elements of the EMS are carefully planned and executed.

- Communications with interested parties is formalized.

TABLE 7-1. Benefits that an Organization Can Expect to Derive from the Implementation of ISO 14001 (*Continued*)

EMS Documentation and Document Control

- All documents critical to managing the EMS are identified, approved, and reviewed and/or revised on a periodic basis so that they remain current.

Operational Control

- Procedures for controlling operations of key environmental activities, including contractor procedures, are documented and current.

Monitoring and Measurement

- Key environmental characteristics are identified, monitored, and measured on a formalized basis.
- Equipment critical to the operation of the EMS is identified, calibration procedures are documented, and calibration records are kept.
- Evaluation of compliance with legal and other requirements is conducted per a documented schedule.

Nonconformance and Corrective/Preventive Action

EMS audit identifies nonconformances within the EMS— examples are missing or outdated procedures, calibration records, training needs, and other system nonconformances. These are investigated and tracked to closure.

Environmental Records

- Environmental records that need to be kept are identified and managed.

EMS Audit

- Complete EMS system audits are conducted on a predetermined schedule and include elements typically not included in other audits, such as employee awareness of the organization's policy and EMS, contractor training, and calibration of equipment.

Table 7-1. Benefits that an Organization Can Expect to Derive from the Implementation of ISO 14001 (*Continued*)

Management Review

- Scope of EMS management review and management involvement is greatly enhanced. Management determines the suitability, adequacy, and effectiveness of the EMS based on EMS audit results, the environmental performance in meeting objectives and targets, changing circumstances, and other EMS results.

- Top management is fully integrated into the maintenance of the EMS.

Table 7-2. Benefits an Organization Can Expect to Derive from Registering to ISO 14001

- Registration validates that employees (from top to bottom), as well as on-site contractors, who did not traditionally see themselves as needing to be involved with the environmental management process are knowledgeable about the organization's EMS. They must demonstrate that they have learned about the environmental policy, environmental aspects and impacts of the organization's of activities, products, and services, as well as the documented objectives and targets. Once aware of the EMS and their part in it, these employees and contractors are more likely to actively contribute the organization's environmental improvements.

- ISO 14001 has the potential to give the organization a marketplace advantage, particularly in Europe and Asia Pacific. In addition, registration demonstrates the organization's quest for environmental leadership.

- Registration validates the organization's top management commitment to sound environmental management through a formalized EMS and to continual improvement of this system.

- Registration validates that the organization's EMS is indeed fully established, implemented, and maintained.

Examples of Continual Improvement Activities*

✓ Additions to and/or strengthening of environmental policy commitments
✓ Additions to objectives and targets
✓ Reductions in air emissions, water discharges, and waste discharges
✓ Reductions in chemical use
✓ Enhanced training activities for those employees who job tasks have the potential to affect the environment
✓ Establishment of a formalized communication plan for internal and external communication about the EMS
✓ Implementation of technology upgrades such as secondary containment, chemical-resistant floor coatings, CFC-free chillers, and aboveground piping
✓ Monitoring and measurement of additional parameters
✓ Improvements to monitoring and measurement tools such as on-line monitors
✓ More frequent tests of emergency preparedness and response procedures
✓ Enhanced document control
✓ Establishment of a field compliance audit program
✓ Increased communications to top management about the EMS
✓ Centralized filing of environmental records
✓ Increased nonhazardous waste recycling rate
✓ Establishment of employee car-pool system

* Continual improvement activities can include improvements to the EMS and/or improvements to environmental performance.

FIGURE 7-1.   Examples of continual improvement activities.

KEYS TO SUCCESS FOR AN EMS AUDITING PROGRAM

FOLLOWING THE AUDIT TRAIL

Throughout this book, and specifically in Chapter 5, there are ample questions for an environmental management system (EMS) auditor to initiate the audit process and address all the requirements of ISO 14001 at all relevant levels of the organization. Almost all of the questions are open ended and begin with who, how, where, when, or what. Answers to these questions typically will lead the auditor on a path that winds its way through the EMS. It is the responsibility of the auditor to tailor his or her audit questions as they follow this path. There are no checklists or question sets that can address all situations that may arise. Thus, an effective auditor is one who can think on his/her feet and ask questions that can lead down the appropriate path to verify closure. This is why trained and experienced EMS auditors are extremely important.

The following is an example of a path one auditor followed over several days during an EMS audit. As shown in the example, the auditor followed a path that began from statements made during the opening meeting and continued during the review of the significant aspects procedure, setting objectives and targets, documented roles and responsibilities, operational control, and monitoring and measurement.

During the opening meeting the EMS management representative stated that the location had taken a very proactive role in identifying and initiating a groundwater corrective action activity as a result of leaks in underground storage tanks, which had occurred approximately 30 to 40 years earlier. Taking note of this, the auditor prepared a few questions to be asked while auditing the aspects and significant aspects procedure. When reviewing the aspects and significant aspects procedures, the auditor looked specifically for the organization's evaluation of groundwater corrective action as an aspect or significant aspect. Here are the questions that the auditor asked the team who performed the evaluation for significant aspects with respect to groundwater corrective action:

- What are the organization's significant aspects?
- How was groundwater corrective action considered in this process?
- Is groundwater corrective action considered a significant aspect?
- What criteria did you use to make this determination?

The auditee indicated that groundwater corrective action was significant and successfully demonstrated how this was evaluated following the organization's procedure.

From this information the auditor prepared a few questions for the upcoming meeting with the people responsible for implementing the organization's procedure for identifying legal and other requirements. During this meeting, the auditor asked the following:

- What are the legal requirements to establish and maintain groundwater corrective action programs?
- How did you identify and communicate these requirements?
- Who is the person responsible for implementing the groundwater program?
- What legally required reports have to be filed as a result of your groundwater corrective action?
- What legally required performance criteria, if any, are established (i.e., objectives and targets)?

Again the auditee successfully responded to the auditor queries. Certain questions, however, came to the auditor's mind about the organization's objectives and targets and documented roles and responsibilities.

When auditing the documented objectives and targets, the auditor asked if objectives and targets had been established for groundwater corrective actions. The auditee responded that there were none currently set. The auditor then asked if this activity was considered when setting objectives and targets. The auditee responded that the activity was considered, but no objectives and targets were set; instead, the organization chose to implement and maintain procedures for operational control of the corrective action. This was an acceptable answer because objectives and targets are not required for all significant aspects as long as there are procedures for operational control.

While reviewing the documented roles and responsibilities, the auditor specifically looked for the inclusion of the person responsible for establishing, implementing, and maintaining the groundwater corrective action plan. In addition, the auditor looked for the person authorized to sign permit applications and reports to legal entities. Based on the document review, the auditor suggested a slight modification to the schedule and asked to speak to the person responsible for establishing, implementing, and maintaining the groundwater correction action plan. When interviewing this person, the auditor asked the following:

- What are your roles and responsibilities with respect to the EMS?
- How were these roles and responsibilities communicated?
- Did you receive any special training to perform your job?
- How were your training needs identified?
- Who has your training records?
- How do you implement and maintain the groundwater corrective action plan?

The auditee responded adequately to questions about roles and responsibilities and training. With respect to implementing and maintaining the groundwater corrective action plan, the audittee responded that there are operational control procedures related to the corrective action plan, which are developed and maintained by a contractor. The auditor then inquired about how the requirements of the corrective action plan were communicated to the contractor and how does the organization verify that the contractor is performing as expected. The auditee produced a detailed scope of work and responded that the contractor provides monthly activity reports, along with sampling and analysis reports. The auditor verified that these reports were available. In addition, the auditor verified that they had been identified as environmental records and that retention times were established. Noting that the activity reports were signed by the contractor and the laboratory reports were signed by the on-site laboratory director, the auditor asked to speak to both people.

The contractor successfully responded to all the auditor questions, which included

- How are requirements for the corrective action plan communicated to you?
- What are the legal requirements pertaining to your activities associated with groundwater corrective action?
- How did you establish operating procedures?
- Who maintains these procedures?
- Are these procedures controlled documents?
- What is your document control procedure?
- Who approved these procedures?
- How do you ensure that obsolete documents are promptly removed from a places of use?
- How do you evaluate the competency of people performing specific tasks to implement these procedures?
- Who does the laboratory analysis?

While interviewing the laboratory director, the auditor noted that samples were analyzed for organic material using a

gas chromatograph. The auditor asked to see instrument calibration records and verification that standards are traceable to
a certified source. The laboratory director produced this material and stated that standards are prepared monthly and stored
in a temperature-controlled refrigerator maintained between
36 and 42°F, according to the legally accepted procedures.
The auditor then asked why the procedures specified specific
refrigerator temperatures, and the lab director responded that
this was necessary to ensure that standards remained stable.

Following the audit trail, the auditor then asked how the
lab director knows that the refrigerator is maintained at the
appropriate temperatures. The lab director produced a procedure that required temperature readings to be recorded daily
and the thermometer to be calibrated annually. The auditor
then asked to see records of the temperature readings and
thermometer calibration records. The laboratory director produced an incomplete log with a note stating temperature readings and calibrations had been suspended (some 3 months
before) because the thermometer in question was broken. The
auditor subsequently wrote a nonconformance for not following the organization's laboratory procedure.

SUBJECTIVE TERMS AND REQUIREMENTS THAT COULD BE INTERPRETED BROADLY

There are several terms within the standard that are very subjective and can be interpreted in several ways. Examples of
these terms (in alphabetical order) and potential interpreta-
tions include

- *Access to legal requirements* The organizations must have a
  procedure to identify and have access to legal and other
  requirements. The term *access* is very broad. Some organizations may have reams of current legal requirements (documents, registers, etc.) available at their location, and others
  may scan monthly update services and access applicable

information via phone calls, wire services, and/or through legal consultants. Further, access to legal requirements can be provided in various ways, including through training, by developing procedures that outline legal requirements, through work instructions and by providing published legal requirements and permits to those who need the information.

- *Changing circumstances*   This can include changes in manufacturing technology, quantity and/or types of parts being manufactured, site mission, number of persons on site, and facility location.

- *Compliance versus conformance*   These two terms are *not* used interchangeably. Compliance refers to meeting legal and/or regulatory requirements; conformance relates to meeting ISO 14001 and/or the organization's internal requirements.

- *Core elements*   These are documents that are integral to the EMS and could include the environmental policy, the EMS manual (should the organization decide to develop one), procedures required by ISO 14001, and other documents determined by the organization to be integral.

- *Documents versus records*   Documents are written requirements that can be changed at any time, such as procedures, work instructions, and inspection forms. Records are objective evidence that an activity or operation has been accomplished, such as monitoring data results, meeting minutes, and completed inspection forms.

- *Key characteristics*   Strictly speaking, key characteristics apply to the operations and activities that can have a significant impact on the environment. Those key characteristics are monitored and measured to track performance, relevant operational controls, and conformance with the organization's environmental objectives and targets. Although it is easy to focus on performance and conformance with environmental objectives and targets, key characteristics related to "relevant operational controls" must not be overlooked. These characteristics can be things such as environmental expenditures, resource allocation, number of agency inspections and resulting non-compliances and/or fines, boundary-level noise, and air emissions from emergency generators.

- *Periodically* The term *periodic* or *periodically* appears in several sections of the ISO 14001 standard, including Section 4.4.4, "Document Control"; Section 4.4.7, "Emergency Preparedness and Response"; Section 4.5.1, "Monitoring and Measurement"; and Section 4.5.4, "Environmental Management System Audits." There is no standard mechanism to determine what this increment is. It must be specified by the organization The term refers to an increment of time specified by the organization (weekly, monthly, quarterly, etc.)

- *Readily retrievable* This basically applies to environmental records and can have varying time frames depending on the record. It may not be necessary to have all records at one's finger tips at all times. For example, an organization may elect to retain environmental property assessment reports for 40 years. Knowing that the need for these records may change over time, *readily retrievable* may mean hours for the first 5 years after the transaction and may mean 5 days after that. On the other hand, if an organization identifies a Material Safety Data Sheet (MSDS) as an environmental record, *readily retrievable* may mean minutes. The important thing is that the records can be found and can be retrieved in a matter of time that is relative to their importance.

- *Relevant* This is one of the two most subjective terms used in the standard—the term *significant* being the other (see below). *Relevant* appears throughout the standard and applies to various requirements including Section 4.2, "Environmental Policy"; Section 4.3.3, "Objectives and Targets"; Section 4.3.4, "Environmental Program(s)"; Section 4.4.2, "Training, Awareness and Competence"; Section 4.4.3, "Communication"; Section 4.4.5, "Document Control"; Section 4.4.6, "Operational Control"; and Section 4.5.1, "Monitoring and Measurement." Every time this word appears in the standard, the organization should go through a process of determining what it means to its system. The meaning of the term will vary depending on the clause in the standard. For example, the organization may determine that *relevant legal requirements* refers to a predetermined list of regulations and permits. *Relevant functions and levels* may be specifically delineated in an organizational chart, and *relevant communication from external interested parties* may include complaints, agency requests, and third-party requests for information about the organization's EMS and environmental performance. The important thing is that the organization determines what *relevant* means with respect to its EMS. The term should not be defined by a registrar or third-party auditor.

- *Significant*   This is another very subjective word that appears throughout the standard. For the most part the use of *significant* is generally associated with environmental aspects and impacts. It must be understood that it is up to the organization to determine what this means within the context of its EMS.

- *Suitable, adequate, effective*   These are terms associated with the responsibility of top management when reviewing and assessing the EMS to determine whether or not it meets the intention and requirements of ISO 14001 and provides expected results. Although these terms are similar and can be found in any standard dictionary, there are subtle differences among them. As with many other terms, it is important that the organization determine what they mean before the management review takes place and also to apply them consistently throughout the EMS.

- *Top management*   The person or persons who ensure that the EMS is suitable, adequate, and effective.

NECESSARY ELEMENTS

There are several "musts," or elements that are necessary for an effective and efficient EMS audit program. These are summarized below:

PLANNING

Planning is perhaps the most important part of the EMS audit. As the adage goes, If you fail to plan, then you plan to fail. Planning includes numerous elements, such as

- Preparing the audit procedure(s) and program
- Preparing the audit schedule and agenda
- Defining roles and responsibilities within the audit process
- Collecting and/or summarizing information for the auditor(s) such as business missions, organization charts, site layout, and other pertinent information

COMMUNICATION

Another key to a successful EMS audit is clear and timely communication. Typically, the audit coordinator is the focal point for communication during the audit process. Clear communication of audit results is, of course, of paramount importance; however, there are typically a myriad of other communications that need to take place during the audit. These include communications about changes in the audit schedule; requests from the auditors for documents or environmental records; communication about what activities, products, and services are (or are not) within the scope of the audit; and communications about logistics (i.e., the need for access to restricted areas, the need for personal protective equipment, which guide accompanies which auditor, lunch arrangements, and other types of day-to-day needs). Thus, the person who is responsible for ensuring that appropriate communications are made should be organized and have good interpersonal skills.

TOP MANAGEMENT COMMITMENT

ISO 14001 requires top management commitment to the organization's EMS, and top management's commitment to the EMS audit process is mandatory for the EMS audit to have meaning. Top management determines the EMS's overall suitability, adequacy, and effectiveness based on these audit results—along with other criteria—but the EMS audit process can also be an important tool for identifying areas for continual improvement. Thus, it is recommended that top management be part of the audit process by attending the opening and closing meeting and scheduling sufficient time with the auditors to ensure that the management review process (Section 4.6 of the ISO 14001 standard) is tested adequately.

PREAUDIT PREPARATION OF THOSE TO BE AUDITED

At a minimum, brief those to be audited on what questions to expect during the audit of their area. If time permits, conduct dry runs with top management, line managers, environmental staff, legal counsel, communications manager, employees, and contractors.

A GOOD ATTITUDE

Having a good attitude about the audit shows the auditors that the organization wants a strong and effective EMS and is able and willing to make changes necessary to align the EMS with the requirements of ISO 14001. The EMS audit is not to be looked upon as contest for proving who is "right"; rather, it should be an open exchange of information for the purpose of improving the EMS.

COMPETENT AUDITORS

If the auditors are not competent to conduct an EMS audit based on their education, training, and experience, the entire audit process is in jeopardy. The organization should define what qualifications are required for an auditor and/or the audit team to be deemed competent. Once defined, the organization should disallow any auditor and/or audit team that doesn't meet these qualifications.

TRACKING NONCONFORMANCES

No matter how thorough an EMS audit is, if nonconformances are not addressed with corrective and preventive actions and if these actions are not tracked to closure, the purpose of the audit is defeated. Typically, the lead auditor consolidates the auditee's response to the nonconformance. In some cases, however, the organization may want to track nonconformances using a software tool or other method of centralized tracking.

ENHANCING THE EMS AUDIT

Once the organization has the "basics" of the EMS audit program in place, enhancing the audit program is an excellent way to continually improve the EMS itself. Examples of what might be considered to enhance the audit program are presented in Figure 8-1. If the organization can make only limited enhancements to its EMS audit process, the authors suggest that it concentrate on enhancing auditor skills and/or the number of qualified auditors. The former makes the audit program more consistent, and the latter makes it more flexible.

THINKING AHEAD: THE POTENTIAL INTEGRATION OF ISO 14001 AND ISO 9001/2 AUDITS

Recognizing that there is considerable overlap between the requirements of ISO 9001 and ISO 14001, the International Organization for Standardization (ISO) is encouraging the two technical committees—TC 176 and TC 207, respectively—to align requirements. Review of compatibility of the two standards is currently underway, and it is anticipated that the next revisions of both will address this issue. Knowing that this is the direction that ISO is considering, organizations should consider integrating the requirements of ISO 9001 and ISO 14001 during the implementation of ISO 14001, because most organizations have an established ISO 9000 program.

Ways that an Organization Might Enhance the EMS Audit Program

✓ Enhance requirements for auditor qualifications through training, education, and/or experience
✓ Expand the pool of qualified EMS auditors
✓ Provide employees and contractors with training on how to effectively answer the auditor's questions
✓ Provide key participants of the audit with a list of "do's" and "don't's" for the audit process
✓ Improve procedure and/or program for root cause analysis of nonconformance, and preventive/corrective action tracking
✓ Review audit results for trends or systemic problems
✓ Increase the frequency of audits in areas that have the most significant nonconformances
✓ Enhance the scope of the audit
✓ Enhance the audit methodology
✓ Develop or purchase a guide for effective auditing

FIGURE 8-1.   Examples of ways to enhance an EMS audit program.

APPENDIX A

EXAMPLE AUDIT REPORT

Date this report was issued: October 6, 1999
Audit Reference Number: XYZ 99

INTRODUCTION

Audit Dates:

EMS audit of XYZ was conducted September 21-23, 1999.

Audit Scope:

Reviewed XYZ organization's EMS as it relates to its activities, products, and services to ensure it meets the requirements of ISO 14001 and other planned arrangements. A complete system audit was performed which tested the XYZ's EMS against all the elements of ISO 14001:96.

Audit Methodology:

The audit methodology used during this audit is defined in *ISO 14001 Implementation Manual*, "Appendix B." The audit process is a sampling process and therefore did not cover every part of XYZ's EMS. The nonconformances found may not address all nonconformances within the EMS, and other internal or third party audits may uncover additional nonconformances. Special attention should be paid to nonconformances found during this audit to determine if they occur in other areas.

References:

ISO 14001, "Environmental Management Systems -- Specification with Guidance for Use," First Edition, 1996-09-01.
Woodside, Aurrichio, and Yturri, *ISO Implementation Guide*, McGraw-Hill, New York, 1998.

Documents / Programs Reviewed:

Documents, programs, and operations reviewed are identified in the agenda and in the auditor's notes provided in Attachment A of this report.

Auditor Qualifications:

The EMS audit team consisted of John Schmidt and Jean Williams, both of whom were independent of XYZ organization's activities, products and services and were, therefore, objective and unbiased. Specific qualifications of the audit team are listed below.

John Schmidt, ABC Consulting

1. Masters in Environmental Engineering
2. Over 15 years experience in environmental field
2. Successfully passed ISO 14001 and ISO 9000 Lead Auditor Training
3. Knowledge of ISO 14001 Standard

Jean Williams, ABC Consulting

1. B.S. in Civil Engineering
2. 10 years experience in environmental field
3. Knowledge of ISO 14001 Standard
4. Experienced in auditing environmental management programs

This report contains:

Agenda
Nonconformance Summary Form
Six (6) Nonconformance Reports
Four (4) Observations

Closure of Nonconformances:

Closure of nonconformances identified in this report will be in conformance with the XYZ's procedure.

Authors' Note: The lead auditor must close nonconformances for registration and surveillance audits. This may also occur with other audits, depending o the agreement between auditor and auditee.

XYZ EMS Audit Agenda

Tuesday, September 21, 1999

Audit Topic	Time	Location	XYZ Personnel
Opening Meeting	8:30-9:00 a.m.	Conference Room C	XYZ Top Management, EMS Management Rep., Environmental Staff, Legal Counsel, Key Area Managers/Reps.
Environmental Policy (4.2)	9:00 - 9:30	Conference Room C	XYZ Top Management, EMS Management Rep..
- Aspects Identification (4.3.1) - Objectives & Targets (4.3.3) - Environmental Mgt. Program (4.3.4)	9:30-10:30 a.m.	Conference Room C	EMS Management Rep.. Environmental Staff
Legal and Other Reqts	10:30 - 11:00 a.m.	Law Library	Env. Attorney
Structure & Responsibility (4.4.1)	11-11:30 a.m.	Environmental Staff Conference Room	EMS Management Rep.
Break - Lunch	11:30-12:30 p.m.		
Monitoring and Measurement (4.5.1) Energy consumption	12:30 - 1 p.m.	Environmental Staff Conference Room	Environmental Staff -- Program engineer for energy
Monitoring and Measurement (4.5.1) Wastewater discharges	1-1:30 p.m.	Environmental Staff Conference Room	Environmental Staff -- Program engineer for wastewater
Monitoring and Measurement (4.5.1) Hazardous waste discharges	1:30-2 p.m.	Environmental Staff Conference Room	Environmental Staff -- Program engineer for chemical/waste management
Monitoring and Measurement (4.5.1) Chemical management/spills	2-2:30 p.m.	Environmental Staff Conference Room	Environmental Staff -- Program engineer for chemical/waste management
Monitoring and Measurement (4.5.1) Air emissions	2:30-3 p.m.	Environmental Staff Conference Room	Environmental Staff -- Program engineer for air emissions
Emergency Planning and Response (4.4.7)	3-3:30 p.m.	Environmental Staff Conference Room	Environmental Staff -- Program engineer for chemical/waste management, Security
Monitoring and Measurement (4.5.1) HAZMAT transport	3:30-4 p.m.	Shipping/receiving office area	Shipping Manager
Auditor work session	4-4:30 p.m.	Conference Room C	N/A
Feedback Meeting w/ Auditors	4:30-5 p.m.	Conference Room C	All who participated in audit activities

XYZ EMS Audit Agenda

Wednesday September 22, 1999

Audit Topic	Time	Location	XYZ Personnel
Meet w/ Auditors to Review Nonconformances	8:00-8:15 a.m.	Conference C	EMS Management Rep.. Environmental Staff. Other Affected Persons
- EMS Audit (4.5.4) - Nonconformance and Corrective/Preventive Action (4.5.2)	8:15-9:00 a.m.	Environmental Staff Conference Room	Environmental Staff -- EMS Audit Coordinator
- Training. Awareness. and Competence (4.4.2) - Communication (4.4.3)	9:00-10:0	Conference Room C	EMS Management Rep.. Training Coordinator: Communications Coordinator
Operational Control Procedures/Records (4.4.6/4.5.3) Manufacturing Area	10:00-11:00	Manufacturing Line	Manager and Personnel from Manufacturing
- EMS Documentation (4.4.4) - Document Control (4.4.5) - Records (4.5.3)	11:00 - 12:15	Environmental Staff Conference Room	EMS Management Rep.. Environmental Staff
Break - Lunch	12:50-1:00 p.m.		
Management Review (4.6)	1:00 - 1:45 p.m.	Top Management's Conference Room	Top Management. EMS Management Rep.
Operational Control Procedures/Records (4.4.6/4.5.3) Chemical/Waste Center	1:45 - 3:00	Chemical/Waste Center	Manager and Personnel from Chemical/Waste Center
Auditor's Meeting	3-4 p.m.	Conference Room C	N/A
Closing Meeting - Audit Summary	4-5 p.m.	Conference Room C	All involved with audit process

Thursday, September 23, 1999

Audit Topic	Time	Location	XYZ Personnel
Meet w/ Auditors to Review Nonconformances	8:00-8:45 a.m.	Conference C	EMS Management Rep. Environmental Staff. Other Affected Persons
Operational Control Procedures/Records (4.4.6/4.5.3) Utility Plants	8:45 - 12:00	Utility Plants A. B. and C	Manager and Personnel from the Utility Plants A. B. and C
Break - Lunch	12:00-1:00 p.m.		
Operational Control Procedures/Records (4.4.6/4.5.3) Contractors	1:00 - 3:00 p.m.	Contractors on Premises Areas: Cafeteria. graphics and copying centers. janitorial services	Appropriate Contractor Managers and Personnel
Auditor's Meeting	3-4 p.m.	Conference Room C	N/A
Closing Meeting - Audit Summary	4-5 p.m.	Conference Room C	All involved with audit process - management

Date: September 21-23, 1999 Audit Team Leader: J. Schmidt Organization: XYZ Ref No: XYZ 99	ENV STAFF	SHIPPING	CHEM CNT	MFG LINE	WWTP	NC REPORT
4.1 General Requirements	X	X				
4.2 Environmental Policy	X	X	X	X	X	
4.3.1 Environmental Aspects	X	X	X	X	X	
4.3.2 Legal and Other Requirements	X					
4.3.3 Objectives and Targets	X		X		X	
4.3.4 Environmental Management Program	X		X		X	
4.4.1 Structure and Responsibility	/	X	X	X	X	1
4.4.2 Training, Awareness and Competence	X	X	X	/	X	1
4.4.3 Communication	X		X	X	X	
4.4.4 Environmental Management System Documentation	X					
4.4.5 Document Control	/	X	X	X	/	1
4.4.6 Operational Control	X		X	X	X	
4.4.7 Emergency Preparedness and Response	X			X		
4.5.1 Monitoring and Measurement	X		X		/	1
4.5.2 Nonconformance and Corrective and Preventive Action	X					
4.5.3 Records	X	/	X	/	/	1
4.5.4 Environmental Management System Audit	X					
4.6 Management Review	/					1
X = Element addressed / = Nonconformance(s) raised within the element	Four (4) observations were noted.		Total NC Reports = 6			

Authors' Note: The EMS audit should test conformance to planned arrangements. Depending on the audit scope, size and structure of the organization, and/or individual auditing style, a nonconformance report form may be written for each nonconformance identified; or, similar nonconformances may be written on a single nonconformance report form. It is important that the auditee understands fully the audit finding and takes appropriate corrective and preventive action. Also, there is no requirement to provide observations. In some cases, the auditee may specifically request that this information be provided within the audit report.

NONCONFORMANCE REPORT

Location: XYZ	Audit Ref No.: XYZ 99
Date: September 21 - 23, 1999	Department/Area: Env Staff
Auditor: Schmidt / Williams	
ISO 14001 Clause Reference: 4.4.1	Major: Minor: X

Requirement: "Roles, responsibility and authorities shall be defined, documented and communicated in order to facilitate effective environmental management"

Nonconformance:

The environmental staff indicated that they received advise and counsel on environmental legal requirements from the XYZ legal staff. Roles, responsibility and authorities of the XYZ legal staff was not documeted.

Signature of Representative Accepting Nonconformance:
Corrective Action Response:

==

Clearance of Nonconformance will follow local procedure(s)

Date Corrective/Preventive Action Completed:
Signature of Representative Approving Corrective/Preventive Action:

NONCONFORMANCE REPORT

Location: XYZ	Audit Ref No.: XYZ 99
Date: September 21 - 23, 1999	Department/Area: MFG Line
Auditor: Schmidt	
ISO 14001 Clause Reference: 4 4.2	Major: Minor: X

Requirement: The organization "shall establish and maintain procedures to make its employees or members at each relevant function and level aware of a) the importance of conformance with the environmental policy and procedures and with the requirements of the environmental management system. ....

Nonconformance:

XYZ's procedure indicated that employees will be made aware of the XYZ environmental policy and their responsibilities in achieving conformance with the environmental policy by the department manager during new employee orientation and/or through department meetings. Several employees on the manufacturing line indicated that they were unaware of the XYZ environmental policy and had not been advised of how their job responsibilities could support the policy.

Signature of Representative Accepting Nonconformance:
Corrective Action Response:

==

Clearance of Nonconformance will follow local procedure(s)

Date Corrective/Preventive Action Completed:
Signature of Representative Approving Corrective/Preventive Action:

NONCONFORMANCE REPORT

Location: XYZ
Date: September 21 - 23, 1999

Audit Ref No.: XYZ 99
Department/Area: Env Staff / WWTP

Auditor: Williams
ISO 14001 Clause Reference: 4.4.5

Major: Minor: X

Requirement: The organization shall establish and maintain procedures for controlling all documents required by this standard...... the procedures will be approved for adequacy by authorized personnel,...... obsolete documents are promptly removed from all points of use...

Nonconformance:

XYZ's procedure indicated that procedures necessary for operational control of the system would be approved for adequacy by the department manager and that obsolete documents are promptly removed from all points of use.

1. Evaluation of procedures maintained by the environmental staff revealed that three of the eighteen procedures sampled had not been approved by the department manager.
2. Evaluation of the operating procedures at the industrial wastewater treatment plant identified two procedures for the treatment of the manufacturing waste stream. When interviewed, an employee indicated that the on-line procedure had been updated a few months ago but the old procedure had not been removed from the procedures manual.

Signature of Representative Accepting Nonconformance:
Corrective Action Response:

==

Clearance of Nonconformance will follow local procedure(s)

Date Corrective/Preventive Action Completed:
Signature of Representative Approving Corrective/Preventive Action:

NONCONFORMANCE REPORT

Location: XYZ
Date: September 21 - 23, 1999
Auditor: Schmidt
ISO 14001 Clause Reference: 4.5.1

Audit Ref No.:   XYZ 99
Department/Area: WWTP

Major: Minor: X

Requirement: "Monitoring equipment shall be calibrated and maintained and records of this process shall be retained according to the organization's procedure"

Nonconformance:

1. XYZ's procedure indicated that the on-line continuous pH monitor would be calibrated at the beginning of each shift (three shift operation). Interviews with employees indicated that this equipment was only calibrated weekly.
2. XYZ's procedure indicated that the truck scale used to weigh hazardous waste sludge transported off site is calibrated annually. There were no records to verify that this scale had been calibrated.

Signature of Representative Accepting Nonconformance:
Corrective Action Response:

===

Clearance of Nonconformance will follow local procedure(s)

Date Corrective/Preventive Action Completed:
Signature of Representative Approving Corrective/Preventive Action:

NONCONFORMANCE REPORT

Location: XYZ
Date: September 21 - 23, 1999
Auditor: Sshmidt / Williams
ISO 14001 Clause Reference: 4.5.3

Audit Ref No.: XYZ 99
Department/Area: Shipping Dept,
MFG Line, WWTP
Major: Minor: X

Requirement: "The organization shall establish and maintain procedures for the identification, maintenance and disposition of environmental records"..

Nonconformance:

1. The procedure for records management established by the shipping department did not identify how environmental records would be disposed of.
2. Employees on the manufacturing line indicated that a critical element for environmental control was the use of environmental assessments (EA's). These EA's are prepared for each manufacturing line and modified when any changes occur. EA's were not identified as environmental records and retention periods had not been established.
3. Training records for employees at the WWTP were not maintained.

Signature of Representative Accepting Nonconformance:
Corrective Action Response:

===

Clearance of Nonconformance will follow local procedure(s)

Date Corrective/Preventive Action Completed:
Signature of Representative Approving Corrective/Preventive Action:

NONCONFORMANCE REPORT

Location: XYZ
Date: September 21 - 23, 1999
Auditor: Williams
ISO 14001 Clause Reference: 4.6

Audit Ref No.:   XYZ 99
Department/Area: Env Staff

Major: Minor: X

Requirement: "The organization's top management shall, at intervals that it determines, review the environmental management system, to ensure its continuing suitability, adequacy and effectiveness"..

Nonconformance:

XYZ's procedure for management review identified the EMS coordinator as the person responsible to determine if the system is suitable, adequate and effective not top management.

Similar statement was made in section 4.4.1 structure and responsibility.

Signature of Representative Accepting Nonconformance:
Corrective Action Response:

===

Clearance of Nonconformance will follow local procedure(s)

Date Corrective/Preventive Action Completed:
Signature of Representative Approving Corrective/Preventive Action:

Auditors' Observations

1. Several minor nonconformances were identified for environmental records management. Additional sampling may have identified this as a systemic issue which could result in a major nonconformance during future EMS audits.

2. Environmental staff engineers individually maintain a list of permits applicable to their area of responsibility. As these lists do not change frequently, the environmental department manager should consider consolidating these lists into one list maintained by a selected environmental staff engineer. This would also be consistent with how environmental department manages its list of applicable legal requirements.

3. XYZ may want to consider expanding the procedure for top management review of the EMS to more accurately reflect what occurs. The procedure currently reflects that EMS reviews with top management occur once a year. In fact, the top management is continually briefed on key characteristics and performance against specific objective and targets through monthly reports. While a complete EMS review must occur annually, this does not have to all take place in one meeting. XYZ's procedure should be written to reflect the partial review of the EMS through these monthly reports.

4. The environmental staff engineers indicated that they worked closely with their procurement department to purchase recycled materials, communicating the organizations environmental requirements to suppliers, and providing information on the organization environmental management system. During the EMS audit this interrelationship was not very clear. XYZ should consider clearly defining the role and responsibilities of the procurement staff to support the environmental policy and the EMS.

APPENDIX B

ISO 14001 AUDITOR TRAINING OVERHEADS

ISO 14001 Auditor Training Overheads

ISO 14001: International Environmental Management System Standard

What is ISO 14001?

- A voluntary international standard developed by the International Organization for Standardization

- It sets the requirements for the establishment of an environmental management system (EMS)

What are the Key Elements of ISO 14001?

- Environmental Policy
- Planning
- Implementation and Operation
- Checking and Corrective Action
- Management Review

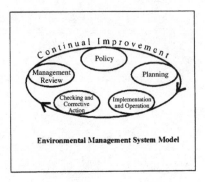

Environmental Management System Model

General Background

ISO 14001 Auditor Training Overheads

Key Benefits of ISO 14001

- Provides integrated approach to environmental management
- Is system dependent and not person dependent
- Demonstrates environmental commitment
- Promotes sound environmental management
- Promotes marketplace leverage

Required Procedures

- Identification of significant environmental aspects
- Identification of legal and other requirements
- Employee awareness of EMS
- Internal and external communication
- Document control
- Operational control
- Emergency preparedness and response

Required Procedures, *Cont.*

- Monitoring and measurement of key characteristics
- Evaluating of legal and regulatory compliance
- Maintaining records
- Handling corrective and preventive action
- EMS audit

Keys to Success

- Management commitment
- Assigned roles and responsibilities
- Implementation plan
- Education, training, and awareness
- Appropriate procedures and documentation
- Audit readiness

General Background

ISO 14001 Auditor Training Overheads

Definition of Environmental Policy - Top management defines policy - Must be relevant to activities, products, and services - Must show commitment to continual improvement - Must show commitment to prevention of pollution	**Definition of Env. Policy, *Cont.*** - Must show commitment to comply with legal and other requirements to which the organization subscribes - Must be communicated to employees - Must be available to the public
Example Environmental Policy Commitments - Fulfill the responsibility of trustee of the environment for this and future generations - To the extent possible, apply practices and control technologies that minimize pollution	**Example Environmental Policy Commitments, *Cont.*** - Comply with legal/regulatory requirements - Strive for improvement of the environmental management system - Prevent pollution through treatment, recycling, and/or reduction

Section 4.2 Environmental Policy

ISO 14001 Auditor Training Overheads

Identification of Environmental Aspects

- Significant environmental aspects of activities, products, and services must be defined
- A significant environmental aspect is one which has a significant environmental impact
- The environmental impact can be adverse or beneficial

Examples of Environmental Aspects

- Planned releases to air, water, and/or land
- Unplanned releases to air, water, and/or land
- Contamination of land
- Consumption of raw materials

Examples of Env. Aspects, *Cont.*

- Consumption of natural resources
- Generation of waste
- Emission of heat
- Waste avoidance
- Creation of wildlife habitat

Significant Environmental Aspects

- These typically have the greatest environmental impact
- They reflect legal and/or regulatory requirements
- They are considered when setting objectives and targets

Section 4.3 Planning

ISO 14001 Auditor Training Overheads

Identify Legal and Other Requirements

- Develop procedures
  for identifying legal and
  other requirements to which
  the organization subscribes

- Ensure relevant employees
  have access to these

Set Objectives and Targets

-These should be measurable
 if at all possible
- These should support the
  environmental policy
- These should consider the
  situation of the organization

Environmental Management Program

- This is a documented program
  to support objectives and targets

- It gives the specifics of "who"
  "when" and "how" the
  objectives and targets are to be
  achieved

Planning

- This section, along with the
  environmental policy, are the
  "heart" of the EMS

- "Fail to plan and plan to fail"

Section 4.3 Planning

ISO 14001 Auditor Training Overheads

Implementation and Operation

- Structure and responsibility
- Training, awareness, and competence
- Communication
- EMS documentation
- Document control
- Operational control
- Emergency preparedness and response

Structure and Responsibility

- Roles, responsibility, and authorities for establishing, implementing, and maintaining the EMS must be defined

- These roles, responsibility, and authorities must be communicated to relevant personnel

Training, Awareness, and Competence

- Employees must be aware of the EMS and its function
- Employees must understand how their work activities support the EMS and the environmental policy
- Employees must be competent in their job functions

Communication

- Information about the significant environmental aspects and the EMS must be communicated internally

- The organization must have a procedure for receiving, documenting, and responding to external communications

Section 4.4 Implementation and Operation

ISO 14001 Auditor Training Overheads

EMS Documentation

- Documents that are core to the EMS must be identified
- Core elements can be in paper or electronic form
- The EMS must provide direction to related documentation

Example Core Elements

- Environmental policy
- EMS manual (if one is developed)
- Procedures required by ISO 14001

Example Related Documentation

- Operating procedures
- Work instructions
- Preventative maintenance schedule
- Equipment calibration procedures
- Inspection forms

Example Related Documentation *Cont.*

- Emergency Plan
- Environmental management program related documents
- Sampling procedures

Section 4.4 Implementation and Operation

ISO 14001 Auditor Training Overheads

Document Control

- There must be a written procedure for document control
- Documents that are essential to the EMS must be controlled

Document Control, *Cont.*

- Documents that are core to the EMS must be controlled
- Documents must be reviewed and revised, as necessary
- Current versions must be available where needed
- Obsolete documents must be removed promptly

Operational Control

- Includes control of operations related to significant environmental aspects and objectives and targets

- Includes control of on-site activities of contractors

- Procedures are required that specify operating criteria

Emergency Preparedness and Response

- Procedures must be in place

- Procedures must be tested, as appropriate

Section 4.4 Implementation and Operation

ISO 14001 Auditor Training Overheads

Monitoring and Measurement

- Key characteristics must be identified and monitored and measured to evaluate environmental performance
- Calibration records for monitoring and measuring equipment must be available
- Compliance with legal requirements must be evaluated

Nonconformance and Corrective/Preventive Action

- Procedures must be in place to handle nonconformance to the EMS

- Roles, responsibilities, and authorities must be assigned

Records

- Environmental records must be easily located and retrievable

- They must be legible, identifiable, and traceable to the activity, product or services

- They must be protected against damage and loss

EMS Audit

- An EMS audit is more comprehensive than a compliance audit

- It should review the system's elements in light of planned arrangements, and should determine whether or not the EMS has been properly implemented and maintained

Section 4.5 Checking and Corrective Action

ISO 14001 Auditor Training Overheads

Management Review

- Must be performed by top management

- EMS is reviewed to ensure its continuing suitability, adequacy, and effectiveness

---

Management Review

- Review process provides information about objectives and targets, monitoring and measurement results, EMS audit results, and other information

---

Management Review

- In light of the information presented, top management makes recommendations for changes to the EMS

- These recommended changes could be changes to policy, objectives and targets, or other elements of the EMS

Management Review

- Management's recommendations for changes should reflect the organization's commitment to continual improvement

- Management review must be documented

Section 4.6 Management Review

ISO 14001 Auditor Training Overheads

Definition of an Audit

An audit is a systematic examination of a process performed by an independent and unbiased person(s) in order to determine if the process complies with planned arrangements

**What are
"Planned Arrangements"**

- Compliance with legal/regulatory requirements
- Adherence to voluntary or industry standards
- Conformance to self-defined requirements
- Conformance to internal instructions and procedures

What is Objective Evidence?

Objective evidence is the "proof" that the organization is doing what it says it is doing

Say what you do; do what you say, **SHOW ME!!**

Purpose of Objective Evidence

The audit process relies on the collection of objective evidence to prove conformance to the requirements of ISO 14001

The Audit Process

ISO 14001 Auditor Training Overheads

Purpose of EMS Audit

To determine if the EMS:

- Meets the intent of ISO 14001
- Conforms to planned arrangements
- Is properly implemented
- Is properly maintained

An EMS Audit Does NOT

Determine if the EMS is suitable, adequate, and effective

This is the responsibility of top management!!!

Types of Audits

First Party Audit: A person within the organization conducts an audit on another part of the organization

Second Party Audit: Typically a customer audit of a supplier

Types of Audits, *Cont.*

Third Party Audit: An audit conducted by an agency or third party registrar

Self Assessment: Evaluation of one's own processes -- Not acceptable as EMS internal audit

The Audit Process

ISO 14001 Auditor Training Overheads

Protocol for an EMS Audit

Pre-audit Activities

- Get a sense of the organization's activities, products, and services
- Understand legal/regulatory requirements pertaining to the organization's business
- Understand the general structure of the organization's EMS

Protocol for an EMS Audit, *Cont.*

Planning

- Select audit team
- Plan agenda
- Develop audit methodology

Protocol for an EMS Audit, *Cont.*

Notification

Notify auditee of EMS Audit at least two weeks in advance; provide audit scope and agenda at that time

---

Protocol for an EMS Audit, *Cont.*

Opening Meeting

- Introduce audit team
- Present scope and agenda
- Discuss ground rules for audit
- Reconfirm that the audit is a sampling process

The Audit Process

ISO 14001 Auditor Training Overheads

**Protocol for an
EMS Audit (cont.)**

Conducting the Audit

- Obtain objective evidence of
  conformance to ISO 14001
- Obtain objective evidence of
  conformance to internal procedures

**Protocol for an
EMS Audit (cont.)**

Closing Meeting

-Auditor relates nonconformances
- Auditee commits to corrective /
  preventive actions and closure
  date(s)

**Protocol for an
EMS Audit (cont.)**

Audit Report

- Team leader gathers information
  and prepares report
- Sample audit report in Appendix
  A of ISO 14001 Auditing Manual

**Requirements of ISO 14001
Pertaining to EMS Audits**

The organization shall establish
and maintain a procedure and
program for EMS audits

The Audit Process

ISO 14001 Auditor Training Overheads

**Requirements of ISO 14001
Pertaining to EMS Audits,** *Cont.*

The audit program and schedule must
be based on the environmental
importance of the activity, and the
audit schedule must cover audit
scope, frequency and
methodology

**Requirements of ISO 14001
Pertaining to EMS Audits,** *Cont.*

Audit results are reported to
top management and are used
as part of top management's
assessment when determining
if the EMS is suitable, adequate,
and effective

Audit Questioning Techniques

Use noncommittal words such as:

- "I see"
- "That is very interesting"
- " I understand"

Purpose: To convey interest in what
the person is saying and to keep
him/her talking

**Audit Questioning
Techniques,** *Cont.*

Use the "Who, what, how,
when, where, and why"
questions

Purpose:
Gather additional facts

The Audit Process

ISO 14001 Auditor Training Overheads

Audit Questioning Techniques, *Cont.*

Restate and reflect the auditee's statement, idea, or feeling

Purpose:
Show you are listening and understand; encourage the auditee to talk

Audit Questioning Techniques, *Cont.*

Summarize

Purpose:
Brings potential noncomformances to the forefront; allows additional information to be provided

Example EMS Audit Questions

See Chapter 5 of
ISO 14001 Auditing
Manual

Putting it All Together

Follow the audit thread

Be flexible and intuitive

See Chapter 8 of ISO 14001 Auditing Manual for example

The Audit Process

APPENDIX C

CORE ELEMENTS OF XYZ'S ENVIRONMENTAL MANAGEMENT SYSTEM

Environmental Policy Statement

The XYZ Company is committed to protection of the environment, and as an environmentally conscious business, XYZ is committed to the following environmental policy objectives:

- Provide sound environmental management practices which will protect the environment for future generations.

- To the extent possible, apply practices and control technologies that minimize pollution.

- Develop and manufacture products that are environmentally friendly and energy efficient whenever possible.

- Comply with applicable regulations.

- Strive to minimize releases to air, water, and land.

- Strive for continual improvement of the environmental management system.

- To the extent possible, use recycled materials throughout the company.

- Prevent pollution through reuse, recycling, and reduction.

To support effective implementation of XYZ's EMS, this policy shall be communicated to all employees.

Issued by Jim O'Brien, XYZ General Manager   01/05/99

Procedure Name: Identifying Significant Environmental Aspects
Document Control Number: EMS 4.3.1
Document Owner/Approver: Mary Smith, Environmental Manager
Date: 01/15/99

The Environmental Manager of XYZ Company initiates the process of identifying environmental aspects of activities, products, and services. The process includes soliciting input from professionals from manufacturing, facilities, procurement, and distribution. The following information is considered:

a) inputs and outputs from routine operations;
b) inputs and outputs from major maintenance and/or turnaround activities; and
c) potential for accidents and emergency situations and their effect on the environment.
d) inputs and outputs from services that could significantly affect the environment

The determination of significant environmental impacts will be based on the numeric process defined in Attachment A of this procedure, with each expert evaluating each aspect. The determination will consider, at a minimum:

a) legal/regulatory requirements pertaining to activities, products, and services;
b) risk to employees and/or neighborhood populations and/or customers;
c) environmental impact frequency;
d) environmental impact;
e) public perception, including customer views.

Environmental aspects which have or can have significant environmental impacts are classified as significant environmental aspects. Significant environmental aspects are documented, reviewed, and updated, as necessary. At a minimum, significant environmental aspects are reviewed annually.

The Environmental Manager communicates the list of XYZ's significant environmental aspects to the XYZ management team and relevant personnel. This communication takes place after the initial identification of significant environmental aspects and whenever there is a change in these. If there are no changes within a calendar year, then the Environmental Manager confirms this to XYZ's management team at the end of the calendar year.

Procedure Name: Identifying Significant Environmental Aspects, *Cont.*
Document Control Number: EMS Procedure 4.3.1

ATTACHMENT A   Environmental Aspect Identification Process

Criteria Used Possible Ratings

1. Legal/Regulatory Requirements (L) -- Is there a legal/regulatory requirement or a permit required?
   0 = There is no legal/regulatory requirement
   3 = There is a legal/regulatory requirement
   5 = There is a permit required

2. Risk (R) - Rate the potential risk to employees and/or neighbor populations.
   1 = Low Risk
   3 = Intermediate Risk
   5 = High Risk

3. Environmental Impact Frequency (F) - Rate the frequency of occurrence.
   1= Low Frequency
   3 = Intermediate Frequency
   5 = High Frequency

4. Environmental Impact (EI) - Classify the impact according the importance.
   1 = Low Importance
   3 = Intermediate Importance
   5 = High Importance

5. Public Perception (P) - Determine the importance of the environmental impact in terms of public perception.
   1 = Low Perception
   3 = Intermediate Perception
   5 = High Perception

Criteria for determining significance of the aspect:

1. Aspect where the sum of values is > or = 15.
2. Aspect where (F+ EI) > 6
3. Aspect where   P = 5

Procedure Name: Identifying and Providing Access to Legal and Other Requirements
Document Control Number: EMS 4.3.2
Document Owner/Approver: Mary Smith, Environmental Manager
Date: 01/07/99

Introduction.

XYZ strives to comply with all legal and other requirements that apply to its operations. Therefore, knowledgeable personnel must continuously monitor new requirements published by local, state and federal authorities to determine applicability for the site. Environmental programs and objectives are modified to incorporate new compliance requirements. Affected personnel must be given access to the requirements, and trained on their implications and impacts. The new requirements and associated changes to affected environmental programs must be added to the internal audit process to assure they have been implemented appropriately.

Requirements and Responsibilities:

Legal Requirements. The Environmental Manager receives a monthly update of regulatory changes from a reputable subscription service. Also he or she scans various websites to access new or proposed regulatory issues that may be of interest. The Environmental Manager is responsible for reviewing new and proposed regulations, making applicability determinations, and printing hard copies of regulations that apply to XYZ. The new requirements are placed in a binder to provide access to others, and be readily available to undergo further study so that modifications to environmental programs can be implemented to assure compliance. In addition, the Environmental Manager forwards new or proposed legal requirements to affected organizations (e.g., manufacturing, facilities, and distribution), for their review. These affected organizations are required to respond to the Environmental Manager within two weeks of their review as to the applicability of the regulations to their operations. The Environmental Manager and the affected organization assess the impact of the new requirements together and develop methods for meeting these requirements.

Regulatory requirements are typically included in work procedures, training materials, and emergency plans.

Other Requirements. XYZ Company does not currently subscribe to other requirements within the scope of its EMS.

XYZ's Environmental Management Program   Page 1 of 2
Document Control Number: EMS 4.3.4
Document Owners: Carol Jenko, April Anderson, and Jon Bulgrat
Document Approver: Mary Smith, Environmental Manager
Date: 02/15/99

Objective: Increase the use of recycled plastic content in new radios.

Targets: (1) By year-end 1999, all radios will have a recycled plastics content of at least 20%; (2) By year-end 2000, all radios will have a recycled plastics content of at least 50%; (3) By year-end 2001, all radios will have a recycled plastics content of at least 70%.

Responsible person: Carol Jenko, Manager, Product Development Staff.

Time frame: Refer to target.

Means for Target (1): Identify companies that recycle plastic; evaluate recycled plastic composition to ensure product suitability (1Q'99). Select company to supply recycled plastic; finalize purchasing agreements; work with development and initiate prototype (2Q'99). Prepare production to use 20% recycled material (3Q'99). Ensure all radios have 20% recycled plastic content (Year-end '99).

Means for Targets (2) and (3): Evaluate feasibility of making these targets; put plan in place for meeting targets (Year-end '99).

Objective: Minimize waste by establishing a return and recycling program for radios.

Targets: (1) By year-end 1999, the organization will establish a plan to accept returned (unwanted) radios from customers (2) By year-end 2000, the organization will recycle 50% of the plastic of all radios returned.

Responsible Person: April Anderson, Product Distribution Manager for Target (1); Stan Kelley, Procurement Manager and Vendor Liaison for Target (2).

Time frame: Refer to target.

Means for Target (1): Select members of working group to develop plan to accept returned radios (2Q '99). Complete written plan (Year-end '99). Implement plan (1Q ' 2000).

Means for Target (2): Review and select vendors who can recycle the plastics from radios (3Q'99); Begin recycling program (Year-end '99). Execute target (Year-end 2000).

XYZ's Environmental Management Program, *Cont.*   Page 2 of 2
Document Control Number: EMS 4.3.4

Objective: Initiate program to identify an environmentally safe deicing material.

Target: None was set for this objective because it is impracticable to do so at this time.

Responsible Person: Jon Bulgrat, Department 452, Environmental and Safety Engineering

Time frame: By year end an environmentally safe deicing material will be identified

Means:

(1) Retain consulting firm with toxicological expertise to perform literature search for environmentally safe materials that have deicing capabilities equal to or better than currently used material. (1Q)

(2) If unsuccessful in finding "off the shelf" alternative, hire chemical laboratory to review reformulation of currently used deicing materials. (2Q-3Q)

(3) Allocate budget as follows: $10,000 for consultant fees; $50,000 for laboratory fees. (1Q)

(4) Hold monthly meetings with the environmental staff, engineering department, and safety department to discuss progress. (Ongoing throughout year)

Document Name: Definition of EMS Roles and Responsibilities at XYZ Company
Document Control No. EMS Document 4.4.1
Document Owner: Mary Smith, Environmental Manager
Approver: Jim O'Brien, XYZ General Manager
Date: 02/19/99

The roles, responsibilities and authorities XYZ's environmental management system are defined as follows:

XYZ General Manager

- Has authority and responsibility to approve and issue XYZ Company's environmental policy
- Responsible for overall environmental compliance
- Has the authority to review and assess the suitability, adequacy, and effectiveness of the EMS
- Has authority to delegate specific responsibility within the EMS

EMS Management Representative

- Has authority, delegated by the General Manager to establish, implement, and maintain the EMS for the XYZ Company
- Has authority to provide the appropriate resources for implementation and maintenance of the EMS
- Responsible for reporting on the performance of the EMS to the General Manager
- Responsible for approving objectives and targets based on recommendation from the EMS core team

Manager of XYZ Environmental Department

- Responsible for identifying training requirements of the environmental staff
- Responsible for approving the XYZ EMS Manual
- Responsible to assign resources to support the EMS

Document Name: Definition of EMS Roles and Responsibilities at XYZ Company, *Cont.*
Document Control No. EMS Document 4.4.1

EMS Coordinator

- Responsible for establishing, implementing, and maintaining the EMS under the guidance of the EMS Management Representative
- Responsible for developing the ISO 14001 awareness training for XYZ employees with respect to the environmental Affairs policy and key elements of the EMS
- Responsible for organization information on the performance of the EMS and presenting it to the EMS Management Representative
- Responsible for coordinating efforts for EMS audits
- Responsible for providing guidance and input to Communication Manager for responding to relevant external communications from interested parties related to the EMS

ISO 14001 EMS Core Team

- Responsible for identifying XYZ's aspects and determining significant aspects
- Responsible for recommending objectives and targets to the EMS Management Representative

Communications Manager

- Responsible and authorized to establish, implement, and maintain sitewide communication vehicles
- Responsible for external communications/interfacing with the public (to receive and document relevant communications from external parties)

Document Name: Definition of EMS Roles and Responsibilities at XYZ Company, *Cont.*
Document Control No. EMS Document 4.4.1

Department Managers

- Responsible for identifying training needs
- Responsible for establishing and maintaining operating procedures necessary for maintaining operation control of department activities
- Responsible for investigating nonconformances and for initiating preventive and corrective action applicable to the department
- Responsible for making employees aware of the affect their actions have or potentially can have on the environment
- Responsible for retaining self-audit records (according to applicable record retention requirements)
- Authorized and responsible for employee evaluations of competency

Environmental Staff

- Responsible for developing, maintaining, and implementing programs to ensure compliance with regulatory requirements, and assessing overall compliance
- Responsible for coordinating monitoring and measurement efforts for key environmental characteristics
- Responsible for external reporting as required by regulation and/or permit conditions
- Responsible for assisting the Environmental Manager to provide guidance and input to Communication Manager for responding to relevant external communications from interested parties related to XYZ's environmental management system
- Responsible for implementing, establishing, and maintaining programs to enable XYZ to meet the defined objectives and targets
- Responsible for tracking progress toward achieving objectives and targets

Security Operations/Emergency Planning Manager

- Responsible for developing and implementing program for emergency preparedness and response to environmental incidents
- Responsible for providing training to employees on emergency response
- Authority to solicit help from external organizations during an emergency

Document Name: Definition of EMS Roles and Responsibilities at XYZ Company, *Cont.*
Document Control No. EMS Document 4.4.1

Emergency First Response Personnel

- Responsible for responding to chemical emergency incidents
- Responsible for advising Security Operations/Emergency Response Manager of corrective action and preventive action taken during incident

Site Contractors and Vendors

- Responsible for adherence to XYZ's environmental policy and operating procedures, that are applicable to their jobs.

XYZ Personnel

- Responsible for adherence to XYZ 's enironmental policy and procedures applicable to their jobs
- Responsible for assisting site in meeting objectives and targets through completing actions as defined by the environmental staff
- Responsible to be aware of the environmental impact of their job

Procedure Name: Procedure for Training, Awareness, and Competence
Document Control Number: EMS 4.4.2
Document Owner/Approver: Mary Smith, Environmental Manager
Date: 03/10/99

Introduction.

XYZ considers training, awareness, and competence paramount to ensuring a strong environmental management system (EMS). Thus, there are requirements for training, awareness and competence of all employees throughout the company whose job functions could impact the environment and the EMS.

Requirements and Responsibilities.

The Environmental Manager is responsible for developing the company's environmental training plan, which meets regulatory requirements and those of the EMS. Further, the Environmental Manager is responsible for developing a training module for EMS awareness training and for scheduling the training classes.

Requirement and Responsibilities for Competence

Job descriptions are written for all employees and describe education, experience, and training needed to perform that job function. In addition, training requirements are defined in department training plans.

Procedure Name: Responding to External Inquiries About Environmental Matters
Document Control Number: EMS 4.4.3A
Document Owner: Mary Smith, Environmental Manager
Document Approver: Jim O'brien, General Manager
Date: 03/29/99

Introduction.

Responding to external inquiries about environmental matters pertaining to XYZ is important to the company. In order to ensure that responses are accurate and consistent responsibilities for answering these external inquiries are defined as follows.

Requirements and Responsibilities.

All external inquiries and/or information requests pertaining to XYZ's EMS are directed to the Environmental Manager, who drafts a response and gets it approved by the General Manager. Inquiries and/or information requests are received through letters or telephone calls and can be from customers, government agency representatives, media representatives, and others who may be interested in XYZ's environmental activities. The Environmental Manager maintains a log of external inquiries, and maintains records of responses to those which might affect XYZ's business or public image.

Procedure Name: Internal Communication About Significant Environmental Aspects and the EMS
Document Control Number: EMS 4.4.3B
Document Owner/Approver: Mary Smith, Environmental Manager
Date: 3/29/99

Introduction.

Internal communication about XYZ's significant environmental aspects and the EMS is necessary in order to maintain a suitable, adequate, and effective EMS. Communications are multi-directional: that is clear lines of communication exist from environmental staff to managers and employees and vice versa.

Requirements and Responsibilities.

The Environmental Manager is responsible for establishing and maintaining clear lines of communication about XYZ's significant aspects and the EMS. This is established and maintained, as follows:

a. The Environmental Manager is responsible for communicating information about the EMS to managers and employees. This information includes, but is not limited to, communications about environmental policy updates, significant environmental aspects, the environmental management program, the EMS audit, nonconformances identified within the EMS, and the results of management reviews. This information is communicated through an annual internal EMS progress report, which is written by the Environmental Operations Officer.

b. The Environmental Manager also communicates to the General Manager about environmental activities or external inquiries that may affect XYZ's business or public image.

c. The Environmental Manager maintains an e-mail address and an internal phone number for receiving questions, information, and other communications from managers and employees.

Procedure Name: EMS Documentation
Document Control Number: EMS 4.4.4
Document Owner/Approver: Mary Smith, Environmental Manager
Date: 04/07/99

Core Elements of XYZ's EMS include the following documents:

- XYZ's Environmental Policy Statement
- EMS 4.3.1 -- Identifying Significant Environmental Aspects
- EMS 4.3.2 -- Identifying and Providing Access to Legal and Other Requirements
- EMS 4.3.4 -- Environmental Management Program
- EMS 4.4.1 -- Definition of EMS Roles and Responsibilities at XYZ Company
- EMS 4.4.2 -- Procedure for Training, Awareness, and Competence
- EMS 4.4.3A -- Responding to External Inquires About Environmental Matters
- EMS 4.4.3B -- Internal Communication About Significant Environmental Aspects
- EMS 4.4.4 -- EMS Documentation
- EMS 4.4.5 -- Document Control
- EMS 4.4.6A -- Procedure for Operational Control
- EMS 4.4.6B -- Procedure for Operational Control of Contractors
- EMS 4.4.7 -- Emergency Preparedness and Response
- EMS 4.5.1A -- Monitoring and Measurement
- EMS 4.5.1B -- Periodically Evaluating Compliance with Environmental Legislation
- EMS 4.5.2 -- Nonconformance and Corrective and Preventive Action
- EMS 4.5.3 -- Records
- EMS 4.5.4A -- Environmental Management System (EMS) Audit
- EMS 4.5.4B -- Environmental Management System (EMS) Audit Program
- EMS 4.5.4C -- XYZ Company EMS Audit Plan

Related Documents that support the core elements of the EMS include:

- XYZ Preventative Maintenance Manual
- XYZ Monitoring and Measurement Program
- Chemical Authorization and Distribution Manual
- Wastewater Treatment Operations Manual
- Utility Plant Procedures Manual
- Equipment Calibration Procedures Manual

Procedure Name: Document Control
Document Control Number: EMS 4.4.5
Document Owner/Approver: Mary Smith, Environmental Manager
Date: 04/07/99

Introduction.

Control of documents that are key to effective management of XYZ's environmental management system (EMS) is necessary to ensure that these documents are handled consistently throughout the corporation.

Requirements and Responsibilities.

Responsibilities for document control are as follows:

a. Core elements of the EMS and those documents that interact with these are listed in XYZ's Control Document Number EMS 4.4.4.

b. The Environmental Manager is responsible for creating and maintaining the procedures and documents defined in the EMS Documentation Hierarchy, and for assuring:

- all controlled documents can be located
- all controlled documents are reviewed at least annually and revised, as necessary
- current versions of the documents are available where needed for effective functioning and management of the EMS
- obsolete documents are removed promptly from all points of use
- permits and other documents retained for legal/knowledge preservation are labeled "Obsolete: For Reference Only"
- controlled documents are legible
- controlled documents are dated, with dates of revision
- maintained in an orderly manner
- retained for one year after they are revised

A master copy of all controlled documents is located in the of office of the Environmental Manager.

Procedure Name: Procedure for Operational Control
Document Control Number: EMS 4.4.6 A
Document Owner/Approver: Mary Smith, Environmental Manager
Date: 04/12/99

Introduction.

XYZ is committed to operational control of its activities, services, and products related to identified significant environmental aspects. Thus, the company has established procedures for operational control which, if not documented and available could lead to deviations from XYZ's environmental policy commitments and objectives and targets.

Requirements and Responsibilities.

Documented procedures for operational control of XYZ's environmental management system are listed in Document Control Number EMS 4.4.4. It is the responsibility of the owner/authorizer for each listed procedure to ensure that the procedure is current, stipulates operating criteria, and is properly communicated.

Procedure Name: Procedure for Operational Control of Contractors
Document Control Number: EMS 4.4.6 B
Document Owner/Approver: Mary Smith, Environmental Manager
Date: 04/12/99

Introduction.

XYZ purchases services from contractors, vendors and suppliers that could have an affect on its environmental management system (EMS). XYZ is committed to ensuring that these service providers understand the importance of supporting the EMS while on XYZ premises.

Requirements and Responsibilities.

The Environmental Manager, in conjunction with the Purchasing/Distribution representative, is responsible for identifying contractors, vendors, and suppliers needed to provide services worldwide related to XYZ's EMS. This includes, but is not limited to, the identification of waste disposal and recycling vendors, suppliers of raw materials, and contractors performing janitorial and maintenance services. As part of the contract, the Purchasing/Distribution representative is responsible to ensure each contractor, vendor, and/or supplier receives the booklet entitled *Contractor Responsibilities While on XYZ Premises*. This document defines requirements to ensure that the contractor works within XYZ's EMS and is aware of how his/her operations and activities can support the Corporation's environmental policy and objectives and targets.

In addition, the Environmental Manager is responsible for evaluating significant environmental aspects and impacts of services requested and for ensuring that these are in line with XYZ's environmental policy and the objectives and targets. Further, the Environmental Manager is responsible for establishing and maintaining a Contractor Audit Checklist. This checklist is to be used to audit contractors to ensure that they meet requirements of the document mentioned above and to ensure that their operations and activities support the EMS.

Procedure Name: Emergency Preparedness and Response
Document Control Number: EMS 4.4.7
Document Owner: Chris Jamieson, Security Manager
Approver: Jim O'Brien, General Manager
Date: 03/22/98

Introduction.

XYZ has identified the potential for accidents and emergency situations. As a result, emergency preparedness and response is a formalized process that addresses prevention, response, and mitigation of these potential emergencies. This process is detailed in the site's *Emergency Action Plan.*

Requirements and Responsibilities.

The requirements and responsibilities pertaining to emergency preparedness and response are as follows:

a. The General Manager has overall responsibility for emergency preparedness and response at XYZ. As such, he or she is responsible for providing resources -- both human and financial -- needed to handle an unplanned release or other emergency that could impact the environment. In addition, the General Manager is responsible for reviewing and approving the *Emergency Action Plan*, designating an Emergency Coordinator, and ensuring that overall prevention, response, and mitigation measures are adequate and effective.

b. The Security Manager has been designated as the Emergency Coordinator. As such, this person is responsible for establishing and maintaining the *Emergency Action Plan*, for coordinating activities during emergency situations, and for testing the *Emergency Action Plan* at least biennially. In addition, the Security Manager, in conjunction with the Environmental Manager, is responsible for assessing, initiating, and documenting corrective and preventive actions following emergency situations.

c. The Security Manager is responsible to maintain a list of employees assigned to job functions for emergency response. In addition, that person is responsible, in conjunction with the Environmental Manager, for coordinating and documenting training to ensure proper response to emergency situations, as well as comply with applicable training requirements. The Environmental Manager is responsible for external communication about accidents and/or incidents, as appropriate.

Procedure Name: Monitoring and Measurement
Document Control Number: EMS 4.5.1A
Document Owner/ Approver: Mary Smith, Environmental Manager
Date: 05/26/98

Introduction.

XYZ has documented a procedural plan for monitoring and measuring its key characteristics. This plan is entitled *XYZ Monitoring and Measurement Program*. In addition, XYZ has identified monitoring equipment necessary to ensure proper operation of the EMS. And, finally, XYZ has a documented procedure for evaluating compliance with environmental legislation and regulations in Document Control Number EMS 4.5.1B.

Requirements and Responsibilities.

The requirements and responsibilities pertaining to monitoring are as follows:

a. The Environmental Manager has overall responsibility for documenting and communicating the *XYZ. Monitoring and Measurement Program.*

b. A list of equipment calibration necessary to ensure proper operation of the EMS is kept by the Environmental Manager. Manufacturer's or other procedures for calibration of this equipment are maintained in the department/area responsible for maintaining the equipment.

c. The Environmental Manager is responsible for establishing and maintaining Document Control Number EMS 4.5.1B.

Procedure Name: Periodically Evaluating Compliance with Environmental Legislation and Regulations
Document Control Number: EMS 4.5.1B
Document Owner: Mary Smith, Environmental Manager
Approver: Jim O'Brien, General Manager
Date: 05/26/98

Introduction.

XYZ is committed to complying with applicable environmental legislation and regulations. Thus, there is a process for periodically evaluating compliance of the manufacturing and assembly activities and operations.

Requirements and Responsibilities.

The Environmental Manager evaluates the activities and operations of XYZ activities, products, services with respect to legal compliance on a semiannually basis, using a compliance checklist that covers all relevant legislation and regulations. Any compliance nonconformance found during the evaluation is noted on the checklist, and corrective and preventive action is taken and documented. Any serious nonconformance identified by the Environmental Manager that could result in fines or in negative publicity is communicated to the General Manager immediately. A summary of the compliance evaluation is prepared by the Environmental Manager and submitted to the General Manager annually.

Procedure Name: Nonconformance and Corrective and Preventive Action
Document Control Number: EMS 4.5.2
Document Owner/Approver: Mary Smith, Environmental Manager
Date: 04/12/99

Introduction.

XYZ is committed to taking corrective and preventive action to mitigate nonconformances identified within the environmental management system (EMS). XYZ has taken such measures over the years when emergency situations have occurred, and is expanding this practice to include nonconformance to all elements of the EMS. Corrective and preventive action will be appropriate to the magnitude of the problem (potential or actual) identified and will be in line with the environmental policy and environmental impact of the nonconformance.

Requirements and Responsibilities.

Requirements and responsibilities for nonconformance and corrective and preventive action at the Corporate and plant site levels are as follows:

The Environmental Manager is responsible for handling any nonconformance to the EMS identified during the EMS audit process or through other means. This person will investigate the nonconformance using root cause analysis and will develop a plan for corrective and preventive action. Should human or financial resources be needed to initiate the plan, these must be authorized by the General Manager of XYZ Once the corrective and preventive action plan is initiated, the Environmental Manager will document and track all such actions to closure.

Procedure Name: Records
Document Control Number: EMS 4.5.3
Document Owner/Approver: Mary Smith, Environmental Manager
Date: 04/06/99

Introduction.

XYZ's environmental records are identified in the XYZ environmental records list kept by the Environmental Manager. Environmental records are legible, identifiable and traceable to the activity, product, or service. They are stored and maintained in such a way as to be readily retrievable and protected against damage, deterioration, or loss.

Requirements and Responsibilities.

Responsibilities for managing records as follows:

The Environmental Manager is responsible for establishing and maintaining the XYZ environmental records list. Record retention times for the various documents are established as follows:

a. Training records are maintained for 5 years
b. Audit results are maintained for 3 years
c. Management reviews are maintained for 3 years
d. Operational data, including monitoring and measurement data, is kept for 5 years
e. Equipment calibration and maintenance records are maintained for 5 years
f. Inspection records are maintained for 3 years
g. Environmental Questionnaires are maintained for 5 years
h. All other records are maintained for 1 year

All records will be disposed as normal trash (with paper recycled when possible), unless otherwise indicated on the record. The Environmental Manager is responsible for ensuring that the records are stored for the prescribed retention time and are maintained and disposed appropriately. The records will be stored in a fireproof cabinet(s) under the control of the Environmental Manager.

Procedure Name: Environmental Management System (EMS) Audit
Document Control Number: EMS 4.5.4A
Document Owner/Approver: Mary Smith, Environmental Manager
Date: 05/18/99

XYZ is committed to assuring that its EMS functions properly. In order to do this, the EMS must be audited by an auditor or team of auditors who are objective and unbiased. Requirements and responsibilities for auditing the EMS at XYZ Company are as follows:

a. The Environmental Manager is responsible for initiating the EMS audit process at XYZ Company. The audit scope, expected frequency, and methodology is determined by this person and is established in the Environmental Management System (EMS) Audit Program, which is defined in the document EMS 4.5.4B. It is the responsibility of the Environmental Manager to be the focal point of interface between XYZ Company and the audit team.

b. The EMS audit team will consist of: XYZ Legal Counsel (lead auditor); a member(s) from the quality auditing team who has had ISO 14001 auditor training; and a member(s) who has three years environmental experience and who is outside the area being audited. It is the responsibility of the lead auditor to conduct the audit in accordance with the XYZ audit methodology and to summarize the findings in an audit report and present it to the President of XYZ Company.

c. Results of the EMS audit will be documented in the XYZ standard audit report format.

d. EMS audit reports will be kept for five (5) years and disposed of as confidential (recycled) waste paper.

Environmental Management System (EMS) Audit Program
Document Control Number: EMS 4.5.4B
Document Owner/Approver: Mary Smith, Environmental Manager
Date: 05/18/99

EMS Audit Scope.

The EMS audit program covers XYZ Companies activities products, and services. All elements of the EMS will be audited, with special emphasis placed on the following:

(1) Awareness and support of the environmental policy and communication of it to employees and contractors
(2) Environmental aspects and significant environmental aspects identification
(3) Understanding of objectives and targets at all relevant levels
(4) Training and awareness methods to ensure that employees understand how their job functions impact the environment and the overall EMS, including the environmental policy and objectives and targets
(5) Communication about the EMS at all relevant levels
(6) Compliance and contractor audits
(7) Adherence to operational procedures
(8) Handling of nonconformance and corrective and preventive action
(9) Management review

EMS Audit Frequency.

A complete EMS audit will be conducted at least annually. Specific elements will be tested more frequently as audit findings warrant.

Environmental Management System (EMS) Audit Program *Cont.*
Document Control Number: EMS 4.5.4B

Results of Audits.

EMS audit results at XYZ Company will be reviewed by the President of XYZ Company to determine if the scope and/or frequency of the audits needs to be changed.

Responsibilities and Requirements for Auditors.

The XYZ Legal Counsel will act as the lead auditor on all EMS audits. This person will follow typical audit protocol. Other audit team members consist of quality auditors who have had ISO 14001 auditor training and a member who has at least three years of environmental experience.

Audit Methodology.

The XYZ audit methodology is defined in *ISO 14001 Implementation Manual*, "Appendix B," McGraw-Hill, 1998.

Document Name: XYZ Company EMS Audit Plan
Document Control No. EMS 4.5.4 C
Document Owner: Dan Johnston, EMS Coordinator
Document Approver: Mary Smith, Environmental Manager
Date: 06/03/99

Planned Audit Dates: September 21, 22, 23, 1999

Auditors Names and Qualifications:
The EMS audit team will consist of Jean Williams and John Schmidt, both of whom are independent of XYZ organization's activities, products and services and are, therefore, objective and unbiased. Specific qualifications of the audit team are listed below.

Jean Williams, ABC Consulting
1. B.S. in Civil Engineering
2. 10 years experience in environmental field
3. Knowledge of ISO 14001 Standard
4. Experienced in auditing environmental management programs

John Schmidt, ABC Consulting
1. Masters in Environmental Engineering
2. Over 15 years experience in environmental field
2. Successfully passed ISO 14001 and ISO 9000 Lead Auditor Training
3. Knowledge of ISO 14001 Standard

Scope: All elements of ISO 14001 with respect to the XYZ environmental management system. This will be a complete system audit.

Audit Process:
Opening Meeting: An opening meeting will be held with the audit team and appropriate XYZ staff. This meeting will include introduction of team members, agenda, audit logistics, and tentative schedule for closing meeting.

Conducting the Audit: The auditors will review documentation and conduct interviews with member of the XYZ as identified in audit agenda. Auditors will examine objective evidence of the selected elements of the EMS. Auditors will obtain acknowledgment of nonconformances.

Closing Meeting: A closing meeting will be held with the audit team and appropriate XYZ staff and management to discuss the results of the audit. The lead auditor must ensure that each nonconformance is understood.

Document Name: XYZ Company EMS Audit Plan
Document Control No. EMS 4.5.4 C

Audit Report: The final audit report will be prepared by the lead auditor. The report is used to document the audit and summarize the findings and will include:

- names of audit team members
- dates of audit
- audit scope
- ISO 14001 elements audited
- nonconformances
- observations

Closure of Nonconformances: Any nonconformance that may arise during the EMS audit will be tracked to closure following the requirements identified in the document EMS 4.5.2.

Areas/Elements to Be Audited

Audit Topic	XYZ Personnel	Comments
Overview of XYZ EMS	EMS Management Rep., Environmental Staff	Present summary of EMS improvements.
Environmental Policy (4.2)	XYZ Top Management, EMS Management Rep.,	Review how XYZ supports the policy.
- Aspects Identification (4.3.1) - Legal & Other Reqts. (4.3.2) - Objectives & Targets (4.3.3) - Environmental Mgt. Program (4.3.4)	EMS Management Rep., Legal Counsel, Environmental Staff	Show updated significant aspects, new objectives and targets, and enhanced environmental management program.
Structure & Responsibility (4.4.1)	EMS Management Rep.	Review communication methods and interview employees
Monitoring and Measurement (4.5.1) Energy consumption	Environmental Staff -- Program engineer for energy	Review energy conservation projects and corresponding data.
Monitoring and Measurement (4.5.1) Wastewater discharges	Environmental Staff -- Program engineer for wastewater	Review training records, calibration records, and operating procedure.
Monitoring and Measurement (4.5.1) Hazardous waste discharges	Environmental Staff -- Program engineer for chemical/waste management	Review training records, operating procedures, and emergency response procedures.
Monitoring and Measurement (4.5.1) Chemical management/spills	Environmental Staff -- Program engineer for chemical/waste management	Review emergency procedures, corrective/preventive actions, and training records.

Document Name: XYZ Company EMS Audit Plan
Document Control No. EMS 4.5.4 C

Areas/Elements to be Audited *Cont.*

Audit Topic	XYZ Personnel	Comments
Monitoring and Measurement (4.5.1) Air emissions	Environmental Staff -- Program engineer for air emissions	Review permits, calibration records, operating procedures, and training records.
Emergency Planning and Response (4.4.7)	Environmental Staff -- Program engineer for chemical/waste management, Security	Review emergency plan, training records, and incident records.
Monitoring and Measurement (4.5.1) HAZMAT transport	Shipping Manager	Review shipping documents and training records.
- EMS Audit (4.5.4) - Nonconformance and Corrective/ Preventive Action (4.5.2)	Environmental Staff -- EMS Audit Coordinator	Review audit plan, methodology, auditor qualification, auditor training records, and audit reports.
- Training, Awareness, and Competence (4.4.2) - Communication (4.4.3)	EMS Management Rep., Training Coordinator; Communications Coordinator	Review training needs, training records, and mechanisms for internal and external communications.
Operational Control Procedures/Records (4.4.6/4.5.3) Manufacturing Area	Manager and Personnel from Manufacturing	Interview employees about how their job can impact the environment.
- EMS Documentation (4.4.4) - Document Control (4.4.5) - Records (4.5.3)	EMS Management Rep., Environmental Staff	Review core elements of EMS and related documents; review document control procedure and test documents.
Management Review (4.6)	Top Management, EMS Management Rep.	Review how top management determines suitability, adequacy, and effectiveness of the EMS.
Operational Control Procedures/Records (4.4.6/4.5.3) Chemical/Waste Center	Manager and Personnel from Chemical/Waste Center	Review training records and operating procedures; interview employees on emergency response actions.
Operational Control Procedures/Records (4.4.6/4.5.3) Utility Plants	Manager and Personnel from the Utility Plants A, B, and C	Review training records, calibration records, and operating procedures.
Operational Control Procedures/Records (4.4.6/4.5.3) Contractors	Contractors on Premises	Interview employees about environmental policy and how their job can impact the environment.

INDEX

ABOUT THE AUTHORS

GAYLE WOODSIDE is program manager on IBM's Corporate Environmental Affairs staff, working on ISO 14001 implementation. She is also the co-author of McGraw-Hill's *ISO 14000 Guide: The New International Environmental Management Standards, ISO 14001 Implementation Manual*, and is the author of *EHS Portable Handbook*. Ms. Woodside is certified as an environmental management systems lead auditor by the Registrar Accreditation Board.

PATRICK AURRICHIO is program manager on IBM's Corporate Environmental Affairs staff, where he is responsible for ISO 14001 worldwide implementation, monitoring and measurement, and EMS training. He also provides technical support for property transactions, environmental assessments, and EMS auditing. With Gayle Woodside and Jeanne Yturri, he is coauthor of *ISO 14001 Implementation Manual*. Mr. Aurrichio is also certified as an environmental management systems lead auditor by the Registrar Accreditation Board.